T0133700

FUNDAMENTALS OF
USER-CENTERED DESIGN
A PRACTICAL APPROACH

FUNDAMENTALS OF
USER-CENTERED DESIGN
A PRACTICAL APPROACH

BRIAN STILL
KATE CRANE

CRC Press
Taylor & Francis Group
Boca Raton London New York

CRC Press is an imprint of the
Taylor & Francis Group, an **informa** business

CRC Press
Taylor & Francis Group
6000 Broken Sound Parkway NW, Suite 300
Boca Raton, FL 33487-2742

ISBN 13: 978-1-4987-6436-0 (hbk)

Visit the Taylor & Francis Web site at
http://www.taylorandfrancis.com

and the CRC Press Web site at
http://www.crcpress.com

Contents

Preface

Products are designed for us every day, such as the websites and mobile applications we use, the cars we drive, and the devices we rely on for work or play. Many theories exist for how to design these items effectively; however, we are often reminded that successful designs are those that allow users to perform tasks, create content, or play games based on their preferences and needs. Still, designers often overlook the user. But user-centered design (UCD) puts the users' needs and wants at the center of the design process from start to finish.

User-centered designers begin collecting feedback before anything is actually designed. They observe and interview users, evaluate the market and competitors, and use this feedback to determine product requirements even before the initial design is prototyped. Instead of waiting until the end of the design process for users to validate a nearly completed product, user-centered designers consult with representative users and let them test multiple prototypes throughout the process. Each time, the design is revised until the product is molded to the users' needs, desires, and situations.

There are many quality books that offer guidance on UCD, but they are limited in scope of content and context. However, our book, *Fundamentals of User-Centered Design: A Practical Approach*, is the first in-depth, comprehensive overview of the key principles, theories, and practices required to employ effective UCD in a variety of design situations.

In 2009, Brian began to teach short courses in user experience (UX) evaluation and UCD, which led to certification through Texas Tech University. At the time, there wasn't a textbook that provided comprehensive content for these students to get a thorough grounding in UX and UCD principles and approaches. Brian created a study workbook for the courses, which he further refined and incorporated in his semester-long university courses. This book is born, in part, from this content.

It also draws on industry knowledge. In addition to Brian's own rich experiences creating and selling UCD products, the ideas, solutions, and stories of UX and UCD professionals, working every day to make products better for users, are critical resources relied on for understanding the practical application of UCD projects.

Fundamentals of User-Centered Design gives its readers not just an understanding of the underpinnings of UCD theory and origins, but also a practical, repeatable process for creating effective UCD. This book includes chapters on user research methods to create actionable user profiles, assessing and creating designs for different use environments, balancing user needs and preferences with client and organizational goals, and building and testing prototypes. UCD principles are demonstrated using timely examples and easy-to-follow descriptions. This book also provides short specialty chapters on UCD for mobile applications, print documents, and hardware. Core chapters also address UCD for international users, working in UCD teams, and tools to help research and produce UCD.

Fundamentals of User-Centered Design is a useful book for students in a range of fields: UX, interaction design, information design, visual design, usability, human factors, human–computer interaction, psychology, computer science, and technical communication. The tone is approachable, and the content is adaptable. Real-world UCD examples (both bad and good) are presented, and takeaways providing thorough summaries to help with comprehension are included for all major chapters. We've also developed accompanying instructor presentation slides and a robust instructor's manual to help students carry out practical research projects. Even practitioners who buy the book will appreciate the thoroughness of the information and the straightforward guidance on how to implement effective UCD for different types of products.

Brian and Kate

Acknowledgments

This book has truly been a collaborative, user-centered project. As chapters were written, students in our classes offered honest, helpful feedback, telling us what they thought was useful to their understanding of user-centered design (UCD) and also what could be made better.

As we developed and revised chapters, Sarah Martin kept the entire process on track. We frankly couldn't have written this book without her. She read every word, she verified references, prepared and fulfilled all copyright permission requests, and managed the creation of what we think is a highly effective, comprehensive instructor's manual that includes supportive PowerPoint slides, lesson plans, and sample assignments. Sarah organized us when we were not, bringing to the project enthusiasm, knowledge, and professionalism. Someday soon we'll be helping her with her books as she has helped us with this one.

Our thanks also go to the publisher, CRC Press, and its senior acquisitions editor, Cindy Carelli, and senior project coordinator, Kari Budyk. Cindy saw the value right away of this book's subject matter and championed it from proposal to completion. Kari patiently and quickly answered every one of our questions about formatting to ensure we delivered our manuscript correctly and on time.

There were other key contributors as well. Nancy Small, Terry Crane, Ana Krahmer, and Greg Gamel helped to collect and evaluate materials. Dr. Brett Oppegaard generously offered in articulate detail the story of the UCD work that went into his award-winning Yellowstone National Park mobile app. And Lara Mandrell's copyediting skills served as a last line of defense for us in making sure we were stylistically consistent and correct. A special nod also goes to Megan Olson, who not only contributed support materials in critical places, but also sent along helpful notes from her careful reading of early stages of the manuscript.

Finally, our thanks go out to our family, friends, and colleagues who supported our work on this project.

For Brian, he owes as always so much to the patience and love of his wife, Amy, who allowed him to steal away on weekends to do extra writing in the midst of an already hectic schedule. Brian's kids, Jack, Olivia, and Abe, deserve thanks if not an apology. There were times Dad spent more time on the book than on them.

For Kate, a special thanks to Jeff and Terry Crane and Ronda Wery for their unwavering encouragement, support, and understanding. My gratitude goes to Cheryl Hall for helping and teaching me to keep all things in perspective. Finally, to my family and friends, who I've nearly ignored for the past year, I'm sorry. In the words of Jon Stewart, "I have heard from multiple sources that [you] are lovely people," and I look forward to becoming reacquainted with you.

Authors

Dr. Brian Still has more than 15 years of industry experience, working in both the private and governmental sectors, as a technology developer, manager, and evaluator. He has directed the TTU Usability Research Lab since 2006, managing a number of user experience (UX) design and testing projects for a wide range of clients and generating a number of publications, including a highly respected collection, *Usability of Complex Systems: Evaluation of User Interaction*, edited with Michael Albers and published by CRC Press in 2011. In the same year, Dr. Still spun off a startup from TTU, EyeGuide, based around eye-tracking technology he helped invent and patent. For the last five years, EyeGuide has made (drawing on sound UX design principles) three different eye-tracking research and control systems, selling them to clients around the world.

Dr. Kate Crane is an assistant professor of technical communication at Eastern Washington University. Her current research focuses on document modality and usability, but her larger interests include usability testing, eye-tracking methodology, and user-centered design (UCD). In addition to her own research, she has worked on UCD and usability projects for EyeGuide and the University of North Texas Libraries' Portal to Texas History. Dr. Crane previously served as the assistant director of the Usability Research Lab at Texas Tech University, where she earned her PhD in technical communication and rhetoric.

1 Introduction to User-Centered Design

In 2009, Jakob Nielsen and his user testing team at the Nielsen Norman Group reported findings from a study on mobile device usability. The results were disappointing. In fact, Nielsen (2009) wrote that watching users "suffering during our sessions reminded us of the very first usability studies we did with traditional websites in 1994. It was that bad" (par. 6).

Most disconcerting and surprising to Nielsen and his team, as well as mobile device designers everywhere, was that users carrying out tasks as part of the 2009 study performed worse than similar users carrying out the same tasks with mobile devices in 2000. In some cases, users of 2009 devices took one or more minutes longer to find something as simple as the local weather forecast than users of 2000 devices.

How could such a result be possible? Those who used mobile devices at the turn of the century remember all too well how frustrating using mobile devices could be, especially when trying to access information from the web. These devices featured tiny screens as well as a tiny touch-tone telephone keypad interface requiring you to choose a letter from a range represented on a number (see Figure 1.1 for a sample of a 2000-era mobile device). Added to difficult input of text was poor network connectivity and interfaces displayed on phone screens clearly designed to display on 1024 × 768 pixel desktop screens. Still, users in 2009 managed to do worse than users in 2000.

Obviously, mobile devices have improved since 2000. In fact, the iPhone was available in 2009 (introduced to the market in 2007) when Nielsen Norman reported on their study. The iPhone and other smart, touchscreen phones performed better in the study than other mobile devices. Nevertheless, despite their better features and quick, widespread adoption, they were not overnight panaceas. The iPhone in 2007, the iPad in 2010, and other extraordinary technology, built with an array of never-before-seen features, have not brought users solutions that deliver mistake-free experiences. One need only look at Google's Wave (Figure 1.2) platform to understand that building state-of-the-art technology does not guarantee a better user experience or market success.

Google Wave was a synchronous (or real-time) communication platform (as opposed to asynchronous communication such as email). The idea was that, rather than cobble together any variety of different technologies to manage projects, share files, instant message, talk, and email, a group of people working together or even just a group of friends or family could use Wave to do everything under one roof.

FIGURE 1.1   2000 era mobile device, the Nokia 3310 "Nokia 3310 blue R7309170 wp."
(Rainer Knäpper, Free Art License; http://artlibre.org/licence/lal/en/.)

Wave was built to give users every tool they might possibly need to share information. One quick glance at the Wave interface certainly demonstrates this:

- Traditional email supplemented by photo profiles of everyone online.
- Built-in search function pulling up returns from key words shared in email
  or in the ongoing chat window in the upper right of the screen.
- An archive of previous communications, file storage, even the capability of
  playing back parts of previous meetings that had been recorded and stored.
  Google Wave had it all (Parr 2009).

With much fanfare and promise, Wave debuted to "an enthusiastic crowd of developers"
in 2009 (Arrington 2010, par. 2). One year later, with only minimal adoption by users,
development stopped. By 2012, Google had officially killed it.

At EyeGuide, the eye-tracking research and control company cofounded in 2011
by Brian Still and Nathan Jahnke, one of the development team's favorite sayings is

FIGURE 1.2   Google Wave "Google Wave in Firefox 3.5 under Mac OS X Leopard." (Jürgen Fenn. Taken on November 28, 2009, Creative Commons Attribution 2.0; https://creativecommons .org/licenses/by/2.0/.)

"making something awesome don't mean using something awesome." With all due respect to Kevin Costner's character in *Field of Dreams*, a fantasy baseball movie from the 1980s, people will not necessarily come if you build it. Just as users struggle with poor technology and don't want to use it, users also struggle with well-designed technology and don't want to use it. If both technologies are built poorly or built well, each has an equal chance of frustrating users or causing them to perform poorly. This results in not only a bad marketing outcome, but often a bad financial outcome as well.

In "Why Software Fails," Robert Charette (2005) writes that software "failures are universally unprejudiced: They happen in every country, to large companies and small; in commercial, nonprofit, and governmental organizations; and without regard to status or reputation. The business and societal costs of these failures…are now well into the billions of dollars a year" (43). One key reason they fail is poor communication between developers and users, which can be seen in the example of Oxford Health Plans. In 1997, Oxford employed a new automated business system without getting feedback on its development or from its users—patients and caregivers. As the business expanded, the system couldn't keep up, resulting in $400 million in uncollected payments and $650 million more in unpaid bills. When public investors learned of Oxford's problems, its stock dropped from $68 dollars a share to $26, a $3.4 billion dollar loss in one day (Parr 2009, 45). Managers and developers initially made no effort to build a system that accommodated user needs, and even once the system was operational, their continued lack of effort to gather feedback from users about how to correct the system ultimately caused it to fail.

Of course, not every company is like Oxford. Charette (2005) mentions Praxis High Integrity Systems as an example of a company that requires "customers be committed...not only financially, but as active participants in" a new system's creation (48). Praxis, according to Charette, focuses a "tremendous amount of time understanding and defining the customer's requirements, and it challenges customers to explain what they want and why. Before a single line of code is written, both the customer and Praxis agree on what is desired, what is feasible, and what risks are involved, given the available resources" (49).

Other organizations are also leading the way in creating development processes and products that engage, proactively, their intended users. Pinterest, Uber, and Airbnb, among others, are examples of organizations that don't focus on making things for users, but rather, as Patrick Stewart (2014) writes, "facilitate the exchange of value between users" (par. 2). At the core of what they do is something Stewart refers to as design thinking.

Design thinking is iterative in nature, but at the heart of how it works is empathy. Designers need to have empathy for a user's experience. As Stewart (2014) notes, Airbnb founder Brian Chesky "rents Airbnb apartments while hosting other users at his apartment" (par. 20) to better "empathize with both renters and hosts and experience the quality of consumer-producer interactions first hand." From this participation, he learns, in an experiential fashion, what customers must do to find a place to stay, what obstacles get in the way of that effort, what tools they use, how and when they engage with those tools, and how those tools do and don't work to aid users in accomplishing their goals. He doesn't let his users design Airbnb for him, but he doesn't design the application without understanding their motives and needs.

Unfortunately, for every Airbnb, there are as many or more that do not subscribe to user-centered design (UCD) thinking. The OpenOffice computer mouse prototype (Figure 1.3), unveiled in 2009, is an example.

Complete with 18 features, including an Analog Xbox 360-style joystick with optional 4-, 8-, and 16-key command modes, 512 kb of its own flash memory, and three different button modes, the OpenOffice mouse represented in form and function the most obvious violation of Steve Krug's (2006) simple but effective first law of usability: "Don't make me think!" (11). Krug explains, "It's the overriding principle—the ultimate tie breaker when deciding whether something works or doesn't..." (11). Krug writes primarily about web design, but this sentiment is true for any product or process people use. To use a product effectively, "it should be self-evident. Obvious" (11).

This truth is not limited to examples of over-designed products or processes. It is easy to pick on things, like the OpenOffice mouse, that fail to be understandable and are not usable for their intended users because they have too many features or functions to comprehend. Certainly, too much utility, or the intrinsic capability of a product or process to complete certain functions or tasks, in a product is a factor in whether users can figure out how to work it or how to use it in particular situations. High-utility interfaces cannot always be avoided. Aircraft cockpit controls, which have been studied since World War II in an effort to make them intuitive to pilots, are an example of high-utility interfaces that are necessary.

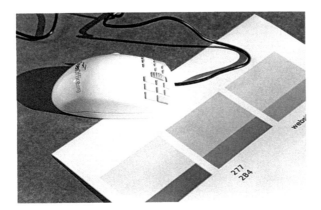

FIGURE 1.3 OpenOffice mouse with 18 built-in functions. (Reprinted with permission from UXcertification.com.)

There are equally unusable low utility interfaces or interfaces that you can only do certain obvious things with while using them. The elevator control images in Figure 1.4a and b illustrate that it is quite possible to make what should be a user-friendly interface unusable. In Figure 1.4a on the left, the elevator control seems simple and straightforward, but the buttons for the floor increase up and to the right in an awkward pattern; Floor 16's button is actually below Floor 5's button. In Figure 1.4b on the right, the floor buttons increase from left to right and include an icon, BR, whose physical location in the building is unclear.

Take a walk around your house (or office or campus) and you're likely to find something designed poorly. In our everyday lives, there are countless examples of things made for us to use but not focused on how we would use them in our real-world experiences. For example, Brian was having lunch at an upscale restaurant. The sink faucets in the bathroom were automatic although Brian initially thought they were broken. Nothing on the faucet itself told Brian how to use it. The soap dispenser made sense, but even when Brian waved his hands beneath the faucet, that action didn't trigger the water to run. In desperation, he used his soapy hands to push down on the faucet, which he noticed was a little taller than it should be. His hunch paid off. The water started. Cold water? Hot water? He couldn't control what he got. Luckily, the temperature wasn't too extreme and he could wash off the soap. However, the water didn't shut off. Brian looked around in a panic. Did he just break the faucet? He reached across and pushed down the faucet in the sink to his right and it came on and started the flow of water. That told Brian the faucet worked by pushing. Eventually, the water stopped, but there were many anxious moments about what might have happened had the water not shut off.

Beyond Brian's mundane example, there are examples of where poor design leads to or exacerbates tragic accidents. The Bhopal, India, gas leak in 1984 and the Chernobyl, Ukraine, nuclear reactor disaster in 1986 are both (Meshkati 1991) prime examples of failures in systems because of a noticeable lack of consideration for human factors. At Chernobyl, despite some engineers' heroic efforts to stop the nuclear reactor's dangerous meltdown, there were no systems in place to allow

(a)

(b)

FIGURE 1.4 Images of confusing elevator controls. (a) Awkward pattern for organizing elevator buttons and (b) what does BR mean? Another example of confusing elevator button labeling. (Reprinted with permission from John Bartholdi, http://www2.isye.gatech .edu/~jjb/misc/elevators/elevators.html.)

operators to carry out actions in emergencies to control the reactor. There weren't alarms to warn them of danger, and by the time the reactor began to fail, there weren't sufficient warnings or tools to prevent the unfolding disaster. To this day, the region around Chernobyl is uninhabitable, and more than 7,000 reported cases of thyroid cancer in children under age 18 have been attributed to the radiation fallout (Nuclear Energy Institute 2015).

Union Carbide's storage plant leaked dangerous gas into Bhopal, India, in December 1984. Almost 3,800 people were killed and another 200,000 injured. Again, although numerous physical and organizational failures led to this disaster, poor design not centered on the needs and capabilities of the plant's operators contributed to the disastrous outcome. Meshkati (1991) notes that plant staff and workers, undertrained and overworked, were responsible for monitoring too many gauges, some of which were often broken. Ironically, many signs informing responsible staff about monitoring plant safety were written in English although the staff predominantly spoke and read Hindi.

In *Set Phasers on Stun*, Casey (1998) catalogues a distressing number of tragic stories stemming from mistakes caused by how users interacted with products. They include the death of a cancer patient repeatedly given the wrong dosage of radiation because the technician had inadvertently typed in an X instead of an E to operate the radiation therapy machine. The crash of an Air France Airbus in 1988, which killed two children and injured many others, is another example. After this crash, and a subsequent one with more fatalities in 1990, Airbus concluded that pilots were too overconfident with how well the automatic landing control system worked in the aircraft. As such, pilots were not sufficiently trained to interpret the system or how to act if the system was providing incorrect data (106).

Endsley and Jones (2004) argue that one reason there is such high operator error in aircraft accidents, for instance, is "not the result of [a] faulty human, but rather the direct result of technology-centered designs that have been ill-suited for promoting high levels of sustained human performance over the wide range of conditions and circumstances called for in the real world. Rather than human error, a more accurate term would be design-induced error" (6). These are examples of technology-centered designers who build the baseball fields and then expect people to come. It's not that users aren't considered in the technology-centered design (TCD) process; rather, they are one of many factors. The primary focus in TCD is to create technology capable of fulfilling certain functional objectives. Only at a later stage of the process, after design and testing for errors in code, are users in the TCD model considered. Unfortunately, then, this feedback comes too late to be constructive. Instead, only surface or cosmetic errors in design can be addressed.

This TCD approach is often called the waterfall approach (Figure 1.5). The product or process being designed falls down toward the users like a waterfall, and each phase of the process is started and completed in succession. The problem with a waterfall is that a lot of water is falling down precipitously, all toward the users, and not any of the water can go back from where it came. This process doesn't allow for much iteration or empathy.

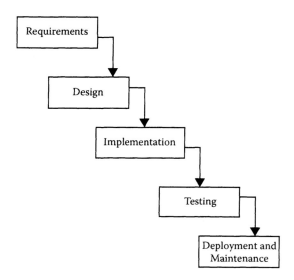

FIGURE 1.5   The waterfall, TCD approach.

The TCD designer needs to answer certain questions:

- What's the deadline?
- What features does the product require?
- What materials are available for constructing the product?
- What's the budget?
- What does the designer believe will make the product successful, aesthetically, functionally, and financially?

TCD designers may try to get a sense of who will use the product through feedback from surveys or focus groups in the beginning, or at the end—when the product is nearly ready to go to the market—when a select group of beta users may give their feedback about the product. However, the actual users and their experiences don't show up anywhere during the design process that makes a product usable for them.

Brian's company, EyeGuide, ironically, built a product born from the classic TCD waterfall approach. In 2012, EyeGuide debuted to the world the first truly affordable assistive technology called EyeGuide Assist. Over 23 million Americans, not to mention millions more elsewhere, have limited or no hand functionality; Brian and his team aspired to rectify this with EyeGuide Assist, which allowed users to control the computer cursor with their eyes. To type or click, Assist enabled users to blink to type or click, to make a sound such as "type" or "click," or to stare at or dwell over a letter on a keyboard or a link on the screen (Figures 1.6 and 1.7).

What EyeGuide promised with all this functionality was also affordability. Many other assistive tools on the market in 2012, especially those that made use of eye-tracking technology, were prohibitively expensive, especially for people with assistive needs on fixed incomes. Having already figured out how to engineer effective,

FIGURE 1.6   EyeGuide Assist in action. (Courtesy of EyeGuide.)

FIGURE 1.7   EyeGuide Assist. (Courtesy of EyeGuide.)

affordable eye tracking with a previous product, EyeGuide felt that Assist offered the right combination of price and functionality.

Debuting at the January 2012 Assistive Technology Industry Association (ATIA) conference in Orlando, Florida, Assist stunned many of those in attendance, including other competitors and disability advocates. It was affordable and could be used to type, even paint, with the eyes. The subsequent press coverage was strong, and EyeGuide launched Assist into the marketplace, taking and fulfilling dozens of orders from all over the world.

It was toward the start of the second quarter in April 2012 that complaints and requests for refunds from Assist customers began to trickle in. Brian and his team worked overtime trying to understand the problems and come up with solutions. Everyone they trained in person still had a unit. One of their more high-profile customers, a former NFL football player diagnosed with amyotrophic lateral sclerosis (ALS), even appeared on an episode of HBO's *Real Sports* with Bryant Gumbel using Assist (Figure 1.8). Still, the refund requests continued, and eventually more than 50% of all users wanted their money back.

Short videos on how to use Assist, increased live and virtual support, and other support tools were implemented to remedy the decreasing sales and increasing refund requests; however, in late 2012, EyeGuide ultimately realized that no amount of support could correct the mistakes that had been designed into Assist, and the product was discontinued. In order to be affordable, Assist needed to cut corners on certain functionality. As Figure 1.8 shows, users or their caregivers had to adjust the camera to fit over the eye, and the LED arm had to be adjusted independently so that the eye could be found in the software and calibrated accurately. Otherwise, the computer mouse, once under Assist's control, wouldn't behave properly. Without head movement compensation built in, users also couldn't move their heads without the risk of losing calibration. Although these limitations were acceptable for researchers carrying out short eye-tracking tests of what users looked at when, for example, buying something online, the ability to put the system on, calibrate, and

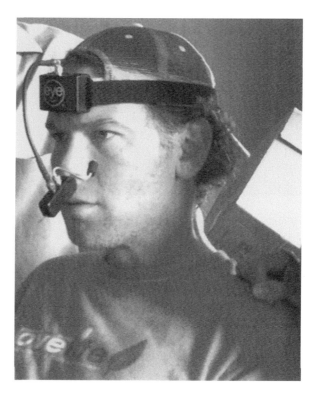

FIGURE 1.8 Former NFL player with ALS using EyeGuide Assist. (Courtesy of EyeGuide.)

quickly function without much additional work for a sustained period of time were all affordances required by assistive users.

No product is cheap enough if it does not work for its intended users. EyeGuide made assumptions about Assist's target user population that were wrong, and only after the product went to market did feedback from users become a part of the process for understanding what they needed and wanted. Technology-centered designers want their products to work. EyeGuide designers wanted to be successful for its investors and for its users. They wanted to make effective, affordable assistive technologies. However, they never bothered to understand what users needed or wanted or how they would use the technology in their everyday lives. They didn't design, in other words, for the user's experience.

Marc Hassenzahl (2014) calls this a tremendous challenge for designers. As the failures of EyeGuide Assist or Google Wave show us, "user experience is not about good industrial design, multi-touch, or fancy interfaces. It is about transcending the material. It is about creating an experience through a device" (Hassenzahl 2014, sec. 1, par. 4). Whereas "the traditional instructional systems design...approach," Baek et al. (2008) write, "is reductionist in nature and that it tends to solve a problem by fragmentation, one stage at a time" (660), design for the user experience is more "subjective, holistic, situated, dynamic, and worthwhile" (Hassenzahl 2014, sec. 3, par. 3). The user experience approach starts with why (Hassenzahl, 2014). The designer does not ask, can I make this or how can I make this? Rather, the designer first asks, why am I going to make this product, and why is it required to assist users in fulfilling an experience?

Donald Norman (2014), responding to Hassenzahl's (2014) emphasis on making design about experience, writes, "Design has moved from its origins of making things look attractive (styling), to making things that fulfill true needs in an effective understandable way (design studies and interactive studies) to the enabling of experiences" (sec. 8, par. 6). It is Norman, writing in 1986, who first proposes how to accomplish this as a means to bridge a natural gap that occurs whenever a designer makes something for a user. According to Norman (1986), a user has personal goals for acting, such as taking money out of an ATM or ordering a shirt from an online store. The problem for designers is that users express these goals in psychological terms. Users want to use products that allow them to easily, effectively, and successfully perform their tasks. However, the system made for them is often problematic because its "mechanisms and states are expressed," according to Norman (1986), "in terms relative to it" (38), not necessarily the user.

Norman (1986) calls these opposed states the "gulfs of execution and evaluation" (38). The user has goals and must interact with the system to execute these goals. The system has a particular interface that needs evaluation. Execution or intention flows from user to system; evaluation or interface, and the means to interpret it, flow from system to user. To put this another way, there is a gap when a user is asked to accomplish a goal with a product. In a perfect world, you would just wish for money and get $40 in your pocket; there would be no need to interface with anything as your wish would be realized by your psychological goals. Unfortunately, this really isn't possible. Therefore, if you're a designer, it is incumbent on you to figure out how to make a product that is as similar and realistic as possible to help the user accomplish goals without too much interpretation or thinking (or at least not any more thinking than is necessary). Obviously, some products, like flying an Airbus aircraft, demand more complicated tasks. Nevertheless, Hassenzahl's (2014) and Krug's (2006) arguments still hold true. To bridge the gap for a better user experience, meaning users can accomplish their goals in a satisfying, effective way, you cannot design from a perspective that emphasizes what the system has to offer and expect the user to adapt to the product. Rather, you should design the system to adapt to the user's psychological, situational, emotional, and intellectual needs.

Therefore, Norman (1986) posits that the best approach to helping designers and users, one that most effectively bridges the gap between evaluation and execution, is through UCD: "User-centered design emphasizes that the purpose of the system is to serve the user, not to use a specific technology, not to be an elegant piece of programming. The needs of the users should dominate the design of the interface, and the needs of the interface should dominate the design of the rest of the system" (61).

Now UCD designers don't stop being designers or dealing with deadlines, budgets, and other constraints just because they focus the design process on the user. They still follow best design practices and adhere to the same requirements found in the waterfall approach. All the waterfall questions matter to them as well. But, and this is key, other questions also matter:

- How would users use this product?
- How do they think about it or similar products?
- How do they interact with this product?
- Where do they interact with it and under what circumstances?

The UCD designer observes and interviews users, evaluates the market and competitors, and uses all of that feedback to determine the requirements needed in a product before it is even prototyped. Instead of waiting until the end of the process for users to validate a nearly completed product, UCD designers consult with representative users throughout the process. Users test out multiple prototypes, and designers make revisions to the design each time. Eventually, the product is molded to the users' needs, desires, and situations.

It doesn't end there either. UCD designers also continue to consult users even after the product is released in order to make improvements and spin off new iterations or products. Many of the applications we use on a daily basis incorporate this attitude of continual design. In fact, application designers often embed protocols into their software that record patterns of use, and they regularly consult user reviews and recruit users for testing in order to develop better versions of their apps.

UCD *doesn't* mean that users are in control of the design process, have the last word on a big decision, or make their opinions count more than others. You also can't make something unique for every individual. Rather, necessary elements of UCD include knowing several key things:

- How the user thinks and interacts with a product
- What the user requires
- What others have done before, both bad and good, to engage users
- What the company wants to accomplish with the product it's making for the intended users

Instead of users evaluating the product at the end of the chain or only evaluating more developed prototypes, UCD values user engagement from the very beginning of the design process. It takes advantage of a variety of methods and techniques that are incorporated into an approach that allows product designers to think about users' needs and visualize how they will use a product from the earliest stages, even before something has been sketched out on paper. If you make things for people to use, you must understand and design for how people will use what you make. Lowgren and Stolterman (2004) have called this "thoughtful design" (5). Basically, thoughtful design "refers to the process that is arranged within existing resource constraints to create, shape, and decide all use-oriented qualities (structural, functional, ethical, and aesthetic) of a digital artifact for one or many clients" (5).

"A key premise of user-centered design," write Endsley and Jones (2004), "is that while a person does not need to perform every task, the person does need to be in control of managing what the systems are doing in order to maintain the situation awareness needed for successful performance across a wide variety of conditions and situations" (11). Essentially, good UCD is not about giving users what they want or making decisions for them. Rather, it is giving them enough control to understand and manage the system in multiple types of situations.

While researching a new application for law enforcement, Brian rode along with a state trooper. He noticed that, on different occasions, particular sounds would prompt the trooper, even in the middle of other tasks, to look at the touch screen monitor positioned to the right of his steering wheel. After the ride-along, Brian asked the trooper about the sounds and their meanings. When everything was in a normal state, there were no unnecessary sounds; thus, the trooper was not distracted and could pay attention to driving his car and viewing other vehicles on the road. However, when a fast beep sound occurred, the trooper quickly looked down at the monitor and noted the position of a flashing dot on a map. He then went back to driving. This beeping sound and corresponding dot indicated that another law enforcement officer had pulled over a vehicle. Should a louder, deeper, longer sound occur, it would notify the trooper that the other law enforcement officer needed assistance. In other words, the sound signaled a change in the situation and because the trooper already made a mental note of the location where the other officer had stopped the vehicle, the trooper could react faster to provide assistance. Further, the loud, deep sound was never used for another purpose. Reserved just for an officer in distress, the trooper would never be confused by its meaning. The interface of the system was designed to aid the user in having control over different situations.

Good UCD is understanding users, designing a product, and doing whatever it takes to make that product work them. Baek et al. (2008) points out that user participation can be weak or strong, in terms of interaction, length, scope, and control. Sometimes users are present at the beginning of the design process, providing constant feedback, trying out prototypes, even contributing to designs. This is strong participation. In other situations, the product's representative users may be recruited to test out prototypes as they are finished, providing more controlled feedback for certain times in the process. This is a weaker form of participation. No one approach, weak, strong, or a combination thereof, is necessarily better or worse than another as long as the participation fulfills project goals and users are involved in some way in the design of the product they will ultimately use.

Along those lines, UCD is a "big tent" approach that draws from a multitude of techniques, methods, and other design concepts. Sometimes you will hear people refer to UCD as more or less another way to describe usability testing. However, they aren't synonymous. We certainly need to administer usability testing of a product as part of a UCD process, but it is just one method that allows you to focus the design process on users. To carry out effective UCD, you may need to understand and apply knowledge of the following:

- Information architecture
- Ergonomics
- Market acceptance
- User preference
- Functionality/utility
- Coding
- Aesthetics

As Pratt and Nunes (2012) write, "well-designed information" has to account for many things (16). One need only watch people try to open doors with unfamiliar handles to understand the importance of everything working well and centering on users' needs. While traveling for work, Brian watched several people get confused by an automated revolving door used to exit the terminal at an airport. There was no guard to warn users that leaving through this exit meant exiting the secured area of the airport. Other than a small sign above the door that read "No Re-Entry," there was no other cue to signal that if users left through the revolving exit door they couldn't come back. Nevertheless, a number of users exited and then attempted to reenter. Likely they were following the crowd of other users who did want to exit and get their baggage. At least one of them was on the phone. Another had her head down and appeared tired.

Should they have seen the sign above the door and acted accordingly? Perhaps, but should the sign or another visual cue have been better placed, given the situation, to guide them in making the correct decision? You design for real people, in the real world, so what you make should reflect an awareness of that. According to Pratt and Nunes (2012), "Understanding who we design for, what they want and need, and the environment in which they will use our designs, is not only a good way to guarantee a successful product, but also a safer, saner world" (16).

UCD is about the entire process from start to finish. For example, if you are a developer, you can follow every protocol to ensure that the code you've implemented has zero defects, yet the product the code powers could still fail because the product wasn't centered adequately on the user's needs. This is also true for aesthetics. Beautiful creations are often rejected because users don't know how to use them, and more than once users have indicated they liked something or preferred that it be made a certain way but didn't buy or use it when it came to the market.

The opposite is also true. A large community of Apple users scoffed at the idea of the iPod, wondering why anyone would want to download individual songs on such a simple device. But Steve Jobs, and Apple, understood that simple design is often the best approach to meeting users' needs. The iPod also granted control directly to the users. The users could control what they downloaded and played. Millions of iPods later, Apple was right, just as it was about the Apple Store, the iPhone, and the iPad. Rather than throw hundreds of things at users in a confusing interface, Apple simplified design and maximized technology, such as touch control, a larger screen for viewing, and a larger keyboard for typing, so that certain affordances were included. They also limited the number of required applications, allowing users to control what was added to their device.

UCD is not focused on just particular aspects of a product's development; instead, it is a thoughtful, comprehensive view of the entire process that puts users' needs front and center. Users' needs are not the only determiner, but they do govern decisions made about aesthetics, architecture, ergonomics, etc. You want these elements of your design done correctly but not in isolation. You design for users who live, work, and play in a real world. UCD focuses on users using things in that world.

The issue then becomes, how do we do it? If we understand what UCD is, how do we implement it in a repeatable, effective way so that it can be successful for you and users, over and over again? This question is what this book attempts to answer. It's important to understand the big picture. We provide that knowledge in this chapter and the next two chapters that look at the history behind UCD and its fundamental governing principles. The bulk of this book, however, aims to give you a comprehensive overview of how to discover, design, and deliver effective UCD.

So let's get started. As Jesse James Garrett (2011) writes, "The concept of user-centered design is very simple: Take the user into account every step of the way as you develop your project" (17). Of course, doing this and saying it are two very different things because "the implications of this simple concept," Garrett continues, "...are surprisingly complex" (17). How should we involve users effectively? Why should users be involved? Are there other ways to design that don't focus so intently on users, and how is UCD different? What events or other factors compelled the creation of UCD as a new approach to design? In this next chapter, we'll focus most on answering this last question as we explore the origins of UCD.

TAKEAWAYS

1. Building state-of-the-art technology does not guarantee a better user experience or market success. Technology built poorly or built well has an equal chance of frustrating users or causing them to perform poorly.
2. Poor communication between developers and users about why, how, and where users use things is a key reason why technologies fail.
3. UCD accommodates user engagement from the very beginning of the design process. Getting feedback or buy-in from users before a product is designed is critical.
4. Design thinking means facilitating the exchange of value between users and having empathy for the user experience.

5. Too much utility or too many features impacts whether users can (or want to) figure out how to use a product.
6. Poor design leads to poor business results and potentially dangerous consequences.
7. Technology-centered design favors fulfilling functional rather than user objectives.
8. No product is cheap enough if it doesn't work for its intended users.
9. A gulf between execution and evaluation means that user goals do not align with how the system evaluates and interprets user input to operate.
10. We should design to make the system adapt to a user's psychological, situational, emotional, and intellectual needs. This trumps a design perspective that emphasizes what the system has to offer above all else and expects users to figure it out.
11. UCD considers a user's mental model: how users might use a product, how they think about it, and how, where, and under what circumstances they interact with it.
12. UCD is not about giving users what they want or making decisions for them. It's about giving them enough control to be able to understand and manage the system they are using, dependent on the situation.

DISCUSSION QUESTIONS

1. Why might a technology, process, or product be more difficult to use now than in the past?
2. Think about EyeGuide's euphemism, "Making something awesome don't mean using something awesome." What's an "awesome" technology or product you have used that turned out not to be so awesome?
3. Envision yourself as a product developer for a new mobile app. What questions might you ask a user before you begin the design process? What types of information do you want to know about them?
4. Think of a technology that you use—by choice or not—on a regular basis. Is it easy to use? If so, why? If it's difficult to use, what would make it easier to use? Discuss how these ease-of-use considerations might be a result of the method, location, or time you use the product.
5. Review your campus or workplace emergency plan for a fire, tornado, or active shooter situation. Are the plans clear and easy to understand and execute? What human factor considerations might help improve user understanding of the procedures? Consider sharing the results with your school or workplace.
6. Think of a step-by-step process that you know how to complete (e.g., getting dressed, cooking, shooting a basketball). How difficult would it be to realize that you made an error and not be able to change anything until the process was complete (waterfall approach)? At what stage would it be easiest to make changes?

REFERENCES

Arrington, M. "Wave Goodbye to Google Wave." *TechCrunch*. August 4, 2010, accessed February 18, 2016. http://techcrunch.com/2010/08/04/wave-goodbye-to-google-wave/.

Baek, E., K. Cagiltay, E. Boling, and T. Frick. "User-Centered Design and Development." In *Handbook of Research on Educational Communications and Technology*, edited by M. Spector, D. Merrill, J. Van Merrienboer, and M. Driscoll, 659–670. New York: Lawrence Earlbaum Associates, 2008.

Casey, S. *Set Phasers on Stun: And Other True Tales of Design, Technology, and Human Error*. Santa Barbara: Aegean, 1998.

Charette, R. "Why Software Fails." *IEEE Spectrum* 42, no. 9 (2005): 42–49.

Endsley, M. and D. Jones. *Designing for Situation Awareness: An Approach to User-Centered Design*. New York: CRC Press, 2004.

Garrett, J. *The Elements of User Experience: User-Centered Design for the Web and Beyond*. Berkeley: New Riders Publishing, 2011.

Hassenzahl, M. "User Experience and Experience Design." In *The Encyclopedia of Human-Computer Interaction*, 2nd ed., edited by M. Soegaard and R. Dam, sec. 3. Interaction Design Foundation, 2014. https://www.interaction-design.org/literature /book/the-encyclopedia-of-human-computer-interaction-2nd-ed/user-experience-and -experience-design.

Krug, S. *Don't Make Me Think*. Berkeley: New Riders Publishing, 2006.

Lowgren, J. and E. Stolterman. *Thoughtful Interaction Design*. Cambridge: MIT Press, 2004.

Meshkati, N. "Human Factors in Large-Scale Technological Systems' Accidents: Three Mile Island, Bhopal, Chernobyl." *Industrial Crisis Quarterly* 5 (1991): 131–154.

Nielsen, J. "Mobile Usability: First Findings." *Nielsen Norman Group*. July 20, 2009, accessed February 18, 2016. https://www.nngroup.com/articles/mobile-usability-first-findings/.

Norman, D. "Cognitive Engineering." In *User-Centered System Design: New Perspectives on Human-Computer Interaction*, edited by D. Norman and S. Draper, 31–61. Hillsdale: Lawrence Erlbaum Associates, 1986.

———. "Commentary by Donald A. Norman." In *The Encyclopedia of Human-Computer Interaction*, 2nd ed., edited by M. Soegaard and R. Dam, sec. 3.8. The Interaction Design Foundation, 2014. https://www.interaction-design.org/literature/book/the-encyclopedia -of-human-computer-interaction-2nd-ed/user-experience-and-experience-design.

Nuclear Energy Institute. "Chernobyl Accident and Its Consequences." *Nuclear Energy Institute*. March 2015, accessed February 18, 2016. http://www.nei.org/master-document-folder /backgrounders/fact-sheets/chernobyl-accident-and-its-consequences.

Parr, B. "Google Wave: A Complete Guide." *Mashable*. May 29, 2009, accessed February 18, 2016. http://mashable.com/2009/05/28/google-wave-guide/#ZmTD5nnbhGqT.

Pratt, A. and J. Nunes. *Interactive Design: An Introduction to the Theory and Application of User-Centered Design*. Beverly: Rockport, 2012.

Stewart, P. "How to Design a Billion Dollar Company." *Applico*. May 15, 2014, accessed February 18, 2016. http://www.applicoinc.com/blog/how-to-design-a-billion-dollar-company/.

2 Origins

User-centered design (UCD), as a framework, derives its approach from a range of disciplines to include techniques and processes for its focus on user needs and wants. Ritter et al.'s (2014) pictorial map in Figure 2.1 provides a useful overview of this dependency.

As shown, UCD is infused by user research, cognitive science (human factors and human computer interaction), ergonomics, and interaction design. Accordingly, these disciplines are reflected in several key questions (Valola 2009):

- How does the user engage with and understand information (interaction design)?
- What is the user capable of understanding, and what can be done to maximize the user's capabilities (cognitive science)?
- What is the best physical environment for the user to use the product, or how can the product be designed physically to make the user's experience successful (ergonomics)?
- What does the user want, where does the user do the work, and what does the user need to use the product successfully (user research)?

The question is how did UCD, as a framework, develop with so many different disciplinary additives to offer a comprehensive approach to design? UCD was first described by Donald Norman in 1986. As influential as Norman's ideas have been since defining a UCD approach three decades ago, his ideas were not immediately adopted. In fact, only in the last decade or so have we seen UCD become an increasingly accepted and implemented approach for designing products for users.

In this chapter, we focus on events from the last 100 years (since the early 20th century) to look at what led Norman to develop a UCD approach and what influenced a more recent rise in the interest of UCD. This time period also corresponds to rapid, modernizing advances in technology that more directly involved users, including those who purchased technological inventions for leisure as much as work. It isn't that people weren't using technology before the turn of the 20th century, nor can we say that users weren't being studied. Wojciech Jastrzębowski coined the term *ergonomics* (Ergoweb 2016), or the study of users and how they can fit better in their environments, in 1857. But we can very easily connect studies done in World War II, for example, to how we study users today. This includes Frank and Lillian Gilbreth, who extrapolated their research of bricklayers in the early 1900s (Price 1989) to create training to speed up the disassembly, cleaning, and reassembly of weapons.

Our overview looks at four thematic periods that advanced chronologically toward our present sense of UCD:

1. Studying human capability and accommodation: The World Wars and after (1914–1976)
2. Making the user relevant: The personal computer (1977–1994)

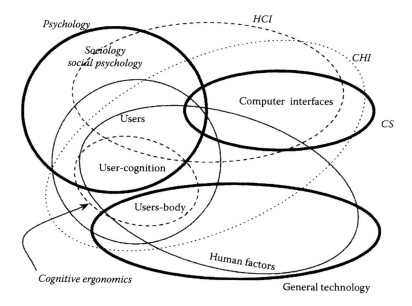

FIGURE 2.1 Pictorial map of fields related to user behavior. (From F. Ritter, G. Baxter, & E. Churchill. *Foundations for Designing User-Centered Systems: What System Designers Need to Know About People.* London: Springer-Verlag, 2014.)

3. Integrating iterative usability into business: The World Wide Web (WWW) (1995–2006)
4. Understanding and designing for the user's ubiquitous computing experience: Mobile, smart devices, and beyond (2007–present)

By somewhat arbitrarily demarcating these periods, we admit that it will be easy to debate, for example, why the first period is so long compared to subsequent periods. Some readers also may immediately see that Norman, arguably the progenitor of UCD, is advocating for the approach 20 years before the beginning of what we consider to be the final UCD-friendly period in 2007. However, it takes a while for new voices to be heard if action is not taken by enough people to constitute a systemic change.

Thomas Kuhn (1970) explains that paradigm shifts, one way of thought revolting against and then replacing a previous way of thought, occur slowly. Testing ideas, those ideas failing, reevaluating the same concept, and then finally seeing something different happens because the prism from which we see it changes and allows us to see it differently. These shifts in science or culture, for instance, are disjunctive, messy, and deliberate. Even Kuhn's notion of paradigm shift, a part of many people's vocabulary, was quickly criticized and not readily adopted.

It is also the case that concepts in other fields may not be known and thus adopted in other areas for some time. UCD can actually be traced to participatory design, which began as a labor movement to democratize the workforce in Sweden and Norway in the 1970s. Douglas Englebart built the first computer mouse in 1963, but not many people knew about it or how to use it until personal computers became popular in the 1980s. Norman wrote about UCD in the late 1980s, but it took the creation and commercialization of the web for his ideas to be employed. For instance, iterative usability testing was developed to make websites more user-friendly and financially viable. Further, reactions to typical usability testing methods and their shortcomings and further advancements in technology, including the smart phone, inspired designers to reexamine Norman's original ideas of the 1980s and adopt them into UCD practices in the 2010s.

As you can see, there are exceptions to our periods and instances of overlap. But rather than fixate on specific start or end times of our contrived periods, we ask that you look at the big events, those that most dramatically and directly impact the everyday user and that characterize each period. For instance, John Bennett published the first paper espousing the value of usability testing for better business returns in 1979. But Nielsen's (1993) popularization of discounted usability testing, connected directly with the explosive growth of the commercial web, had the biggest direct impact on the most users. We build on both the shoulders of giants and little phenomena. Our purpose for recounting history is intended to demonstrate the efforts that occurred during each period to gather feedback from and about users during these big events. Retracing these events allows you to see how studying the way people think, act, and use things has advanced somewhat steadily with one thing building upon another. UCD has become what it is today because of these building blocks and contributions of a rich array of different inputs.

2.1 STUDYING HUMAN CAPABILITY AND ACCOMMODATION: THE WORLD WARS AND AFTER (1914–1976)

For as long as history records, people have built things meant to be used. Four hundred thousand years ago, humans made the first spear; 40,000 years ago, they made their first plow and other agricultural tools (Telier 2011). As technological capabilities improved (from stone to bronze to steel, from steam power to gas turbine to nuclear reactor, and from paper mail to telegraph to email and beyond), we have created products that have immediately faced the task of teaching people how to use those creations. Not until the last 100 years or so did we really give any thought about how user needs or wants should be factored in when creating something.

If you had a tool to use, it was likely that you made the tool yourself, were given the tool and didn't have a choice but to use it, or the tool wasn't difficult to master. In other words, if you were a farmer, you had to use a tool like a plow to do a job. The plow was not especially complicated to use. The work was repetitive and not intellectually challenging. No special research could be done that would lead to new methods for making you better or faster at plowing a field. You plowed because you had to plow. In some ways, you were an expert with the tool because you were a farmer, and as such you learned over time to integrate the plow into your work so that you could use it better than an average person who rarely or never used it. Because that average person would never use the plow because he or she wasn't a farmer, it was not necessary to study how to make the plow more user-centered in its design or provide instructions for someone who might want to switch careers and take up plowing. Moreover, a business didn't have to worry about making the plow pleasing to the farmer or selling the farmer on how easy the plow was to use. It was a plow, the farmer needed to plow, so the farmer bought the plow.

If you were a factory worker and couldn't do your job, the business would fire you and hire another worker to replace you. Even as industrialization made technologies for mass production, there was not much thought given to how a worker's performance could be improved with better instructions or interfaces for the technology they used.

Over time, things changed. Agencies such as the Bureau of Mines and the National Safety Council improved worker safety dramatically in the early part of the 20th century (Aldrich 2001), which meant workers couldn't be forced to work unending shifts in unsafe conditions. Regular work days and weeks required manufacturers to treat their workers better and, in return, made workers more productive while they worked. The old solution, which made employees work until the job was done, was no longer an option. Henry Ford, the father of the world's first affordable car, was likely the first industrial revolution developer who understood there was money to be made in combining efficiency, simplicity, and affordability. Ford didn't let users be designers. If he had, so the quote (although likely not true) attributed to him goes, people wouldn't have asked for cars, they would have asked for "faster horses." Instead, Ford studied users, both his workers and potential consumers, and concluded that he could make more money if relatively unskilled people (not engineers) contributed to the manufacturing of cars on an assembly line and, in turn, he could sell those cars at a price that even his workers could afford. UCD designers don't listen to everything users want. They solve problems for users by making user needs and wants the active focus of their designs. Ford got the result he wanted, and users and consumers got a product they wanted.

Improving user efficiency in the early 20th century didn't just mean better workers. Catering to the new middle class, who were capable of buying new technologies, meant having to explain how inventions like an electric refrigerator or an in-home telephone could be easy to use and life-changing. In a 1936 advertisement, Frigidaire heralded its new refrigerator (Figure 2.2) as having more usability: "wider, roomier, handier to use."

A 1930s era Bell telephone (Figure 2.3) had a built-in manual that provided instructions to the user on proper telephone etiquette. Most people who would buy and install this telephone had never used one before. Bell understood it needed to provide guidance on not only how to make and receive phone calls but also what to say on those phone calls.

Other new consumer technologies that nonexperts used required designers to find ways to integrate them into the home. This shift also meant manufacturers had to convince users that they could use these products. Consumers' satisfied use translated into improved business.

However, war, more than business, was the pivotal impetus for a sustained, comprehensive effort to study users and account for their capabilities through training or design. World War I (1914–1918) required governments to arm millions of mostly uneducated, untrained people in modern warfare. Unlike wars of the past when men were given a uniform and a pike or musket and sent off into battle, men were trained to clean and use repeating rifles, machine guns, howitzers, tanks, and even airplanes.

FIGURE 2.2 Frigidaire advertisement from 1936 *Palm Beach Post* highlighting new refrigerator as having "more usability." (Courtesy of Dr. Brian Still.)

There was no process in place to assess what job soldiers could do or what training would help them do their job. As a remedy for this lack of understanding of human capability, countries began to conduct standardized testing of recruits. Waterson (2011) notes that, in the United States between 1917 and 1918, "approximately two million recruits…were tested in batches of up to 500 a time using intelligence tests" (1113).

Aside from newly recruited people in uniform, many people were thrust into industry to take over jobs left by soldiers. A large percentage of these workers were women who were even less educated than men because of educational standards and societal norms of the day. They worked extra shifts, often in dangerous munitions factories, to keep the military supplied with weapons and ammunition. Worker loss from resignation or injury was high as were strikes protesting poor conditions. This prompted the United Kingdom to establish an Industrial Fatigue Board to study and improve conditions (Waterson 2011), which included more training and better equipment interfaces.

FIGURE 2.3   1930s Bell telephone. (Courtesy of Dr. Brian Still.)

As challenging as World War I was, World War II's experience far overshadows it. There is a direct line from research conducted during this war that continued after it. In many cases, entire fields (such as the now widely practiced human factors field) owe their beginnings to the efforts of scientists in World War II who carefully studied people in new ways.

At the start of World War II in 1939, the U.S. Army was the 17th largest in the world with approximately 200,000 soldiers on active duty (Nelson II 1993). By the end of the war, more than 8 million people were in uniform. Counting all branches of the armed forces, more than 12 million soldiers had been mobilized (Nelson II 1993). Still, more than 30% of eligible soldiers were denied enlistment because of medical conditions (Karpinos 1960), including malnourishment from the Great Depression, and a larger number of people lacked a high school diploma (U.S. Department of Education 1993). Yet, without this education, those who passed the necessary physical tests became soldiers, sailors, marines, and airmen. These soldiers were suddenly in a machine-powered military, not the horse-powered one of their fathers or grandfathers.

The airplane was an example of the more modern military. Although airplanes were present in World War I, the airplane of World War II represented an environment full of user interaction issues, such as cockpit controls, not previously experienced. The cockpit of a Spitfire (Figure 2.4), a fighter airplane used by the British Royal Air Force, illustrates the complicated decision making that confronted pilots flying it.

During the early stages of World War II, airplane failures during training and later in combat were initially attributed to physiological factors, such as a lack of oxygen or pilot fatigue. But early experimentation that asked pilots to carry out their tasks in simulations showed that "poor design of controls and instrumentation were also to blame" for airplane crashes (Waterson 2011, 1119). As Broadbent (1980) wrote, "The technology was fine, but it seemed to be badly matched to human beings" (44). As the Spitfire image shows, the heads-up display in some cockpits could be overwhelming. Seating and other physical characteristics of the plane made it difficult to operate effectively, especially under dangerous conditions in combat. Alphonse Chapanis, then a lieutenant in the U.S. Army and later a leader in the field of ergonomics (the study of people's efficiency in their environment), examined cockpits and experimented with ways to reduce pilot error by constructing more intuitive layouts.

Chapanis studied a series of B-17 bomber crashes in 1943 (Lavietes 2002). He learned that the controls used to operate the B-17's landing gear and flaps were the same shape and design. This caused pilots to confuse them during landing: Instead of deploying flaps to aid in landing, they retracted the landing gear. In response, Chapanis proposed different shapes for the knobs of the landing gear and flap controls, enabling pilots to see and feel the difference. This technique, called shape coding (Woodson and Conover 1964), helped reduce training crashes. Ultimately, the

FIGURE 2.4   Spitfire cockpit controls ("Spitfire Controls," public domain, https://commons .wikimedia.org/wiki/File:Spitifire_controls.jpg).

work of Chapanis and others contributed to improvements in how service members were trained and how weaponry was made to be used more easily.

The focus on human factors and ergonomics research from World War II continued through the 1950s and 1960s. Millions of retuning soldiers from a modernized battlefield returned to an equally modernized workplace. Researchers in the 1950s wanted to know more about human capability. Psychologist George Miller studied human memory, positing in 1956 a now famous notion that people are capable of retaining and using just a few pieces of information, seven plus or minus two specifically, in their working memory (Miller 1956). Alan Baddeley and others in the 1960s and 1970s expanded on the idea of working memory to show that everyone has a limited amount of space in the working memory "which is divided between storage and control processing demands" (Baddeley and Hitch 1974, 76). If too much demand is placed on storage, the user can't process as much information and vice versa.

Researchers like Baddeley and Miller attempted to understand how people think and how to improve the thinking process and capabilities. Industry also wanted to maximize its employees' productivity, convinced by the effectiveness of such work during the preceding war. Throughout the 1960s and 1970s, fields such as human computer interaction (HCI), human factors, and ergonomics flourished. They still flourish today. Their emphasis on exploring user performance, how we think, what we desire, and how machines or physical environments make performance better are all crucial components of UCD today.

But as influential as these fields are, they are relatively systems-centered. In other words, the emphasis for studying user ability or tendency was to make better systems. Knowledge of users in TCD (see more about technology-centered design in Chapter 1) is included in system requirements at the outset of every project, but users are not active players in that process. Therefore, the typical human is a subject of study to make the system better for the typical human.

As the post–World War II years moved on, a hot war was replaced by a cold war against communism. Advancements in computers, satellites, and rocketry led to the development of Arpanet (which would later become the Internet, a communication failsafe for the government and researchers in case of nuclear war) and the space race. As the United States and the Soviet Union battled to put a better rocket and multiple flight missions into space, researchers focused more effort than ever on understanding the limits of what a human could or could not do. The goal was a perfect system. Once the system was perfect, humans would be educated to use it properly. If humans couldn't be trusted to operate the system, then researchers pursued alternatives. In fact, NASA elected to send a monkey into space before a human.

Culturally, the response was interesting. Americans worshiped astronauts because they had *The Right Stuff* (Wolfe 1979), and no one seemed smarter than a rocket scientist. At the same time, we were concerned about technology's potential consequences, such as a nuclear holocaust destroying the earth or losing control to machines that became smarter every day. Stanley Kubrik's popular 1969 film, *2001: A Space Odyssey*, portrayed a future dominated by intelligent, maniacal machines. Through human study, better systems helped Americans win two wars, go to the moon, and keep communism at bay. However, the systems-centered focus also created anxiety.

2.2   MAKING THE USER RELEVANT:
THE PERSONAL COMPUTER (1977–1994)

Foucault (1978) argues that where there is power there is resistance to it. Reactions to systems thinking began to take shape and have an impact. The American counterculture movement of the late 1960s, backlash against government overreach in the Vietnam War, fallout from President Nixon's impeachment and resignation, and other similar events in the United States and abroad signaled rebellion against establishment or system-oriented thinking. Kristen Nygaard, influenced by labor movements in the 1970s in Scandinavia to advocate for workers to have more of a voice, pioneered participatory design. Nygaard initiated "collaborations between computer science researchers and union workers to create a Norwegian national agreement to ensure the rights of unions regarding the design and use of technology in the workplace" (Baek et al. 2008, 662). This new method took hold and spread to the United States intent on making users "human actors" with fewer political motives (662). Participatory design methodology moved us from technology- to user-focused design with which user needs were understood and integrated, the environment was considered, and the user was an active participant in the design.

While participatory design was underway in Northern Europe, underground computer development in the United States was also beginning. This development was another response to the establishment's engineering practices since World War II and eventually led to the creation of one of the world's most profitable companies ever: Apple®. Steve Jobs, Apple's founder, was an original member of the Homebrew Computer Club. Homebrew, founded in 1975 (Wozniak n.d.), was a place where people interested in building computers and related peripherals, such as circuit boards or software, gathered, talked about the work they were attempting to do, and helped each other solve problems they encountered. Homebrew members saw themselves as an alternative movement dedicated to building technologies that challenged previous approaches for building them.

Jobs wanted to empower people. He also saw a viable business model by making technology people wanted and could use for work or leisure. Apple's now famous 1984 Super Bowl commercial for its new desktop computer, the Macintosh, is a perfect example of Jobs' antipathy for the establishment. In the commercial, people are dressed in drab clothing, sitting with vacant stares and mouths agape, watching an old man spout ideology from a large screen in front of them. These are the users of the pre-personal computer age: They have limited choices for their computer needs but are brainwashed into believing that they have no other options. Then, a healthy, beautiful woman being chased by authorities quickly runs down the aisle, stops, and hurls a heavy mace into the screen, destroying it in a pyrotechnic flash. A new voice is heard saying that thanks to Apple, 1984 will not be like Orwell's *1984*. There will be a revolution in technology, one that puts users first by giving them personal computers.

Jobs was much like Henry Ford. He didn't give users what they said they wanted because they would have asked for the same computers they used. As Norman previously noted, users want what they think can be easily realized. Sometimes, what users think is possible doesn't match what technology is actually capable of delivering. Jobs' now famous quote "people don't know what they want until we show them" (Aguado 2015, xviii) illustrates this misconception. Often, technology cannot deliver. But sometimes technology can deliver more than expected. Jobs had a conceptual design vision that he worked to render through the physical product and accompanying explanations of it (such as marketing and documentation) so that users understood the design and would want to buy it. Desire, comfort, experience, and knowledge were all built into one product. Jobs wanted owning an Apple Macintosh to be cool, but he knew people needed to use it to perform tasks they needed or wanted to perform without a steep learning curve.

The Apple Macintosh wasn't the first personal computer (PC). The three preceding it were referred to as the "1977 Trinity" (Williams 1985): the Apple II, the Tandy TRS-80, and the Commodore PET (often considered the first commercially successful PC). Because they all debuted for the home market in 1977, this is where we mark the beginning of a new period that saw both technological development and changes in how users were treated as consumers and participants. For the first time, users had access to tools they made or worked with, and the information from other workers was shared via networks with personal computers. This shift was transformational. Previously, people worked at typewriters. If a company had a computer, it was a mainframe computer and as big as an entire large room, like the IBM 360 seen in Figure 2.5.

If workers connected to the mainframe, they did so through dummy terminals, which were devices without their own processors that could only input or display data. Often, only the mainframe operator could carry out inputs by feeding rectangular paper punch cards into the mainframe; these punch cards contained commands for the computer to execute. Brian had just started college in 1986 when students, during the mainframe heyday, had to stand in line in the school gym and

FIGURE 2.5 IBM System/360 mainframe. (Courtesy of Erik Pitti, March 28, 2008. Creative Commons Attribution 2.0.)

wait their turn to receive a punch card for a class they wanted to take. After receiving these punch cards, students would then deliver them to the mainframe room to be uploaded. In fact, this is why Brian took Japanese instead of Spanish in college. Every user will try to beat the system, and Brian, with a last name of Still, had no chance of taking Spanish because the line of students, all in alphabetical order, ran from inside the gym to outside of it, full of seemingly hundreds waiting to take Spanish. Brian noticed a line with no one and asked what class it was. Learning it was for Japanese language instruction, he promptly started the first of three semesters in the subject to fulfill his language requirement.

The PC, especially one connected with others in a local network that was linked to other networks, changed the workplace almost overnight. Its impact was not limited to work. In fact, its entry into the home space required computer and software application manufacturers to reconsider how they sold what they made. In World War II, airplane cockpits were reconfigured to be more efficient, and pilots were trained to use them. Even the PC in the workplace was about efficiency overall, and workers still had to complete some level of training, although not as much as the pilot, to use them to do their jobs. But businesses couldn't make people at home train more than they wanted to use a PC. For the first time, businesses not only had to convince people that PCs were easy to use, but they also had to make them easy to use; otherwise, people would complain about the product or wouldn't buy it at all. What users thought, wanted, and needed mattered more with the introduction of the consumer PC and derivative software.

To deliver products people wanted and could use, researchers drew from ergonomics, human factors, HCI, and other fields to develop a methodology, usability testing, that constructively incorporated user feedback in the design process. The International Organization for Standardization in ISO9241-11 defines usability as "the extent to which a product can be used by specified users to achieve specific goals in a specified context of use with effectiveness, efficiency, and satisfaction" (UsabilityNet 1998). Therefore, usability testing evaluates users' use of the product. What we know about usability testing today likely originates from Whiteside, Bennett, and Holtzblatt's work, *Usability Engineering* (1988). Responding to industry and researcher demands for synthesizing user testing into something effective and repeatable, *Usability Engineering* served as a catalyst for other instrumental studies in usability testing and subsequent publications. These studies called for recruiting representative users to try out products, and researchers recorded and analyzed their performance. During testing, users only carried out realistic tasks. The point of testing was not to make the product fail, but to discover how the product could be improved so representative users would find it efficient, effective, and satisfying. Think-aloud protocol, developed by Ericsson and Simon (1980), to gather verbal feedback became bedrock to usability testing. Further, John Brooke's System Usability Scale (SUS), created in 1986, to measure satisfaction was integrated so all requirements of the International Organization for Standardization (ISO) for usability were measured in the typical test. (Chapter 8 provides more discussion of usability testing procedures.)

An entire usability evaluation profession emerged. In 1991, the Usability Professionals Association (UPA) was founded as a community and advocacy group for usability consultants and researchers. In 1993, Joe Dumas and Ginny Redish published *A Practical Guide to Usability Testing*. Jeff Rubin followed in 1994 with his book, *The Handbook of Usability Testing*. Usability testing found itself a necessary methodology as developments in new technology emerged, such as the birth of the WWW in 1991 and its commercialization in 1993. Usability testing's efforts to establish itself through practice and scholarship in the 1980s and early 1990s were rewarded by demands for even more testing with the explosive growth of the WWW. By the end of 1994, there were more than 10,000 websites around the world and, just two years later, more than 650,000 (Gray 1996). As transformational as the PC was, nothing matched the historical growth of the WWW. Not even the telephone or electricity experienced this rapid adoption (DeGusta 2012).

Many merchants began to sell products online as the WWW boomed: Amazon was introduced in 1996 and eBay in 1998. Tens of thousands more ecommerce websites were developed as the WWW transitioned from a research to commercial space. Unfortunately for most people trying to shop online in the late 1990s, the only thing as prevalent as virtual shopping options was really bad virtual shopping experiences. Fortunately, usability testing not only offered a means to make PCs and their software user-friendly, but also turned to fixing website interface problems where people increasingly spent their money.

2.3 INTEGRATING ITERATIVE USABILITY INTO BUSINESS: THE WWW (1995–2006)

We focus the start of this period in 1995 for four reasons:

- The beginning of the WWW's exponential growth, including commercial sites, began to flourish.
- Netscape Navigator (in December 1994) and Internet Explorer, the first full-featured browsers, debuted. Because they worked differently than each other in how they displayed web pages, web designers had to use different code, which caused serious interface compatibility issues for users.
- The release of Windows 95's operating system, the first operating system (OS) built with consumers in mind, featured a task bar with a menu, start button, and other features designed to enhance user interaction.
- The debut of Jakob Nielsen's Useit.com website offered advice to experts and novices on how to build better products, especially websites, for users.

It wasn't that Nielsen's work, such as *Usability Engineering* (1993), was the first to discuss the commercial applicability of usability testing. As early as 1979, others, such as Bennett, made the business case for usability testing. Further, Randolph Bias and Deborah Mayhew (1994) wrote an important book, *Cost Justifying Usability*, which laid out how to quantify usability results into dollars saved or earned. However, Nielsen and his Useit.com website popularized usability and made usability accessible to a nonexpert audience. In turn, Nielsen devised an approach, discount usability testing, which was easy enough for nonexperts to employ. Discount usability testing could be prepared and conducted quickly. Web developers and designers tasked with coding their websites, making them compatible with different browsers, and maintaining servers and databases, found this approach to be a more inviting way to gather user feedback.

To this historical point, most usability testing functioned like other testing requirements in a TCD, waterfall approach. This testing did involve real users carrying out real tasks, but the evaluation came at the end of the process; whatever was discovered couldn't be reincorporated usefully into the design. Test reporting was tedious, and reporting requirements were even worse, often asking researchers to compile endless pages of details that no one wanted to read or didn't know what to do with if they did. Heuristic evaluations were a workaround for reporting, usually providing a checklist of best practices based on previous research. The designer could then apply a heuristic to a website and identify and fix problems by comparing, for example, how many mouse clicks (or inputs) it took to find information as opposed to the heuristic's note on how many clicks it should take. Although heuristics are fast and generally reliable, they are not the same thing as testing real users.

Nielsen's discount testing gave guidance to designers like Brian who did want to test real users. A webmaster for a government entity's website in 1996, Brian was the only one responsible for creating and maintaining the website. Before Microsoft FrontPage, Adobe Dreamweaver, active server pages, JavaScript, or easy-to-configure-and-connect-to-the-back-end databases, webmasters like Brian spent almost every day experimenting, failing, rebooting, and then doing it all over again. Servers crashed if too many users were connected simultaneously. Because most users connected through dial-up connections, large images caused a lot of frustration. Creating forms that could dynamically process users' inputs and submit them to a database for storage or to some other script for processing was the most challenging task. Help for webmasters mostly came from other struggling webmasters and developers. But Nielsen's Useit.com was online and accessible, provided easy-to-understand information and guidance, and presented a reliable voice about integrating usability testing into development for webmasters and designers who were contributing to the dot.com boom.

The discount usability testing approach was straightforward: Don't wait until the end of development to test. Problems caught early in development could be easily fixed and cost less. Testing iteratively, or multiple times during development (Nielsen advocated testing at least three times), meant that, at successive stages, a website could be tested, errors found and fixed, then another set of problems could be found and fixed, and so on. Drawing on Virzi (1992) as well as his own research, Nielsen concluded that roughly five users were needed to find about 80% of the problems that affected the largest part of the user population. With three rounds of testing and 15 users, most major problems would be discovered at different stages of the website's fidelity. A website test cycle may evolve, using five users each round, like this: test when the website was being developed; test again when it was halfway through development, which may take the form of a wireframe because it looked functional but wasn't live; then test when the site was close to finished and ready for viewing.

Elaborate preparation wasn't necessary for this formative, nonhypothetical testing. A large, randomized user population wasn't necessary because website designers weren't trying to prove anything about a population's tendencies or about a theory they had about a particular kind of interface working. They just wanted to find usability problems that impacted the most people. Because users that had the most problems weren't experts and novices, or first-time users stopped being novices soon after they used a website, the population to recruit and use for testing were what Carol Barnum (2002) calls perpetual intermediates: people who use only the features they deem necessary to complete their goal but do not attempt to learn the site's intricacies or get better at using the site. These users have a goal they want to accomplish, and they don't want to encounter technical issues, such as poor shopping cart checkout. Rather than defining an in-depth persona, Nielsen encouraged designers to find five intermediate users based on their experience and motivation. Users would perform a few real-world tasks using the website while their actions were recorded and other data (such as time on task, errors, mouse clicks, verbal feedback, think-aloud protocol) were collected.

For Brian, it seemed a logical, accomplishable way to test users, and he and thousands upon thousands of designers relied heavily on discount usability testing. Reports on test results were clear and actionable, usually delivered through PowerPoint, highlighting what was wrong (and how we knew it was wrong from user data collected during testing) and recommendations to fix it. In some ways, discount usability testing felt akin to a thermostat, a metaphor that Genov (2005) in the very first issue of the *Journal of Usability Studies* drew on to describe it.

For a thermostat to function effectively, it needs to monitor the environment. It takes in data about the room temperature, measures this input comparatively to what the temperature should be, and then makes adjustments to the temperature so that the environment is regulated. If a thermostat only measured the temperature once a day, it wouldn't have a good sense of whether, for example, a room's temperature was what it should be. By measuring it constantly or at different key stages, the temperature can be more effectively regulated.

Iterative usability testing, functioning like the thermostat, seeks a more constant flow of feedback from users during the design process. Following from control systems theory, according to Genov (2005), "the initial state of the system," or the first prototype, is evaluated; the results of the test or "input" are compared to goals or "reference value," and if there are any differences "between the current state and desired state of no problems" (24) the system is fixed, a new output (or prototype) is created, and the system is tested again. Therefore, iterative usability testing treated like a control system or thermostat, therefore, is "a form of feedback, which is most effective and resource-efficient…" (Genov 2005, 25).

Certainly, many believed so and still believe so. Usability testing should be done iteratively and as rapidly as possible and the results integrated into a system's redesign. This is especially true in agile software development environments in which designers are working quickly to create and commit programming code so that improvements and solutions to problems are integrated as soon as possible for release into updated system versions. If usability testing cannot keep up with such sprints, it doesn't have any value.

Iterative usability testing is not UCD. It is one methodology UCD relies upon. However, in the mid-2000s, we learned there were problems associated with typical iterative usability testing that made us consider improvements to the method or opt for other approaches. These problems and their solutions moved us into the current period in which complex user experiences are explored and accounted for in a UCD process.

2.4 UNDERSTANDING AND DESIGNING FOR THE USER'S UBIQUITOUS COMPUTING EXPERIENCE: MOBILE, WEARABLE SMART DEVICES, AND BEYOND (2007–)

We mark this period's beginning as 2007 because this was the year Apple introduced the iPhone. Like those before it, this period's starting point was marked by a seminal event that impacted the most everyday users. The iPhone's debut, followed by mobile tablets, dramatically changed how and where users use technology. The desktop PC and laptop were no longer the only way to go online. Users, especially those in countries with limited broadband infrastructure to support Internet connectivity or where computers were too costly, flocked to the mobile smart device for their information needs. People downloaded dozens of applications on their mobile devices for navigation, dining out, playing games, sharing photos, and even for talking to other people. These mobile devices were also becoming more frequently used for work in inventive ways, including displaying X-rays for physicians.

As information access changed, there were good and bad outcomes. Rumblings of discontent about the disadvantages and limitations of usability testing as it was being practiced started before the iPhone's introduction to the market. Lane Becker (2004) declared that 90% of usability testing was useless, not because it didn't collect user data, but because of how it was used as part of the overall design process. Molich et al. (2004) noted that a comparative evaluation of different usability tests conducted on the same application generated different typical user tasks, found different problems, and offered different recommendations. Usability testing such as this was exposed for its lack of context. Testing was done for testing's sake but was not given a purposeful role in the design process.

A growing recognition of the disconnect between what users did in real life compared to what designers thought or designed for came back to Norman's observations in 1986. User testing, even iteratively, didn't account for the fact that in most instances the product hadn't been built with sufficient feedback or knowledge of where users would use it or how they would want or need to use it. Usability testing improved the product, but it didn't help tell designers if the product was the right one for users given their needs and wants.

It was in this same time period that Brian worked with a client to test a smoking cessation kiosk at a public health facility in Texas (Still 2011). For this project, Brian's team was tasked with helping the client understand why the kiosk was received unfavorably by both staff and customers. Several usability tests were conducted, and all major user problems with the interface had been found and corrected. Still, the staff at the facility told the client that customer requests for information from the staff about smoking cessation had gone up, not down, after the kiosks had been installed. Further, customers asked questions about how to use the kiosks although they were designed to be self-explanatory to reduce staff time.

After investigating the situation, including interviewing customers, conducting on-site visits, and performing follow-up usability testing, Brian and his research team concluded that the kiosks had not been designed to account for visitors who were computer illiterate (couldn't use computers and had no experience with touchscreens), functionally illiterate (couldn't read or write), or did not read or speak English. While testing in a usability lab, problems with the interface had been found and corrected. However, the client never bothered to consider who would use the kiosks. In addition, the introduction of something so sophisticated in a relatively unsophisticated office environment felt a little bit like dropping an obelisk into the Stone Age, to reference again Kubrik's *2001: A Space Odyssey*. The kiosks weren't made to fit the intended environment or user community.

Researchers and designers wanting more user involvement in the design process recognized that present means for gathering and using it were wanting. An approach to user involvement in design couldn't start at the end or only involve surveys or interviews at the beginning. It had to be comprehensive, taking into account a greater range of possible factors influencing use and design. Finally, it had to consider that usability problems were not finite. Problems appeared, could change, or could suddenly become problems depending on a number of factors, including dynamic user environments.

The world was very different. We were becoming even more distracted. Because the technology was better and faster, it could give us results faster, and we wanted results as fast as possible. As our access to information became more mobile, any of the number of tasks we were performing and managing intellectually inhibited our ability to make sense of and effectively use faster and larger amounts of information bombarding our memories 24 hours a day, seven days a week. Same brains plus more distraction equals problems with using information.

Think of it this way: Before using mobile devices for the web and all its information, the web was something we visited, as we would visit other things (Figure 2.6). Our everyday lives were separate unless we sat down at work or home and connected to the web through our PC or laptop.

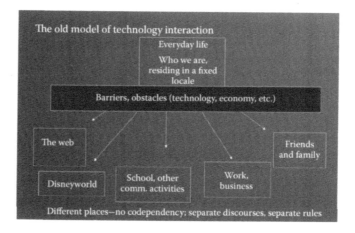

FIGURE 2.6   Old model of technology interaction.

The mobile smartphone intertwined the web with our everyday lives, and that connected everything else we did. Media and information became an "always on" resource (Figure 2.7), and we began to consume it at ever-increasing levels. From 2008 to 2013, the average American's media/information consumption increased from 11 to 14 hours a day (Short 2015). In 2015, it was 15.5 hours a day.

Complexity, not complication, fostered by ubiquitous technology and information, brought about a new set of problems for users and those who developed products for them. As Nielsen's (2009) mobile device study illustrated, users were performing tasks worse using better technology than when using inferior technology almost a decade prior.

Albers (2004) defined complex systems as follows:

- [The] total system is more than just the sum of the parts (a complete description cannot be given).
- Multiple paths to a solution exist.
- Information requirements to answer a question cannot be predefined.
- Effect on the overall system of change cannot be predicted (a nonlinear response).
- The system is open, which makes it difficult to break out and study a section in isolation (16).

Complexity refers to information overload. Flying an aircraft is complicated because there's a lot to learn. However, a pilot can be trained to fly a plane. Searching for directions on your smart device is simple. A user uses Google or another app to find directions. However, if environmental distractions are added in (e.g., bad weather for the pilot, noise or movement for the mobile device user), then tasks become more

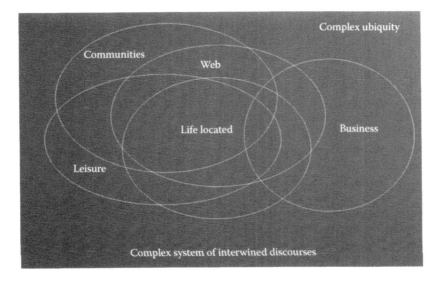

FIGURE 2.7   New model of interaction.

difficult and complex to carry out. Therefore, a complex system is open-ended, multidimensional, and dynamic. Complexity is about lack of control and situational awareness. Distraction, which results from this complexity, causes frustration, poor decision making, and poor performance.

There are fascinating studies about distraction and memory (Morgan 1917; Banbury et al. 2001; Hyman et al. 2010). Hyman et al. (2010) conducted a test comparing whether cell phone users would be more distracted than non-cell phone users to notice a clown on a unicycle go by a university campus. Overwhelmingly, the cell phone users ignored the clown more than non-cell phone users. This distraction has been termed *digital deadwalkers* when people pay more attention to information on their mobile device than they do the reality around them. Halsey (2015) reported that "in 2005, 256 pedestrians [were] injured while using phones" but that this number "grew sixfold by 2010" (par. 10).

Brian and his team conducted another study of an online reference tool to support writing. To make the environment as realistic as possible, his team allowed test participants to make phone calls, text, watch television, use other software on the computer, listen to music, or do whatever they typically did while they worked. After testing was completed and time on task was analyzed, they determined that, on average, every user *spent only 18 seconds (30%) of each minute actually engaged in a specific task.* Had they forgone this field research and conducted a typical usability test, they would not have revealed this user behavior. The team might have found problems with the interface of the online reference tool, but they would not have understood the environment users occupied when using the product—an environment that could make simple task performance more challenging and complex.

Essentially, this lack of situational awareness is a problem that plagues laboratory usability tests, and, over the last decade, usability researchers have realized that we need to move away from tradition during the design process. As Redish (2007) notes, traditional testing "is too short, too 'small task'-based, and not context-rich enough to handle the long, complex, and differing scenarios that typify the work situations that these complex information systems must satisfy" (106). By visiting students in their actual use environments, Brian's team could develop more representative testing scenarios. Rather than assigning students specific tasks to complete, they were free to use the product to complete normal work: edit a paper, complete a homework assignment, or study for an exam. Importantly, students were welcome to utilize other technologies or engage in other tasks that they normally would while conducting their work; students were free to make phone calls, text, check social media, play music, etc. Students were engaged in a representative context—full of representative distractions—that they normally encountered as they worked.

This ecological method of observing users in relation to their complex use system (Still 2011) didn't yield easily quantifiable, timed results. However, it did uncover important relationships about users and their complex use scenarios. Specifically, although the tertiary technologies and activities students engaged in while they conducted work were sometimes distracting, they also aided students in completing their work. The environmental "noise" also served as a means for task completion (i.e., students sought assignment clarification via text messaging and chatted on social media). The distractions that delayed efficiency also worked as resources.

Based on this knowledge, Brian's team was able to provide a concrete, contextually informed UCD recommendation to the client: include messaging and community forum capabilities to their product. This recommendation may not have been considered under typical testing circumstances.

When you design for users, you must first figure out who they are and how they interact with the world. You must observe the complex system they operate in. You want to find out how, when, where, and why they do what they do. Only then can you capitalize on contextual insights that are limited in methods like traditional usability testing or interviews alone. UCD starts and grows with users; the construction of a product should always be based on knowledge about users and the complex systems they use products in.

Ultimately, as we continue navigating this new mobile, complex world, we need UCD because the typical design processes don't work in isolation. UCD is about situated, dynamic, contextualized design focused on the users' needs and wants. It builds into its approach the theoretical concepts of other user-concerned fields (e.g., cognitive sciences, ergonomics) because it wants to create the fullest picture of the user experience as possible. The movement to address the complexity of ubiquitous computing and replace limited applicability of traditional usability testing with a user experience (UX) methodology is the last influential force at work during this period of time to impact UCD's increased popularity. Leaders like Garrett, who first wrote *The Elements of User Experience* (2011), ask us to take what Tharon Howard (2015) calls a more constructionist, less accommodationist approach to incorporating user feedback. According to Howard, we are now less concerned about making products work for users and more focused on constructing a better situation that "allows them [users]…to successfully experience the interface" (Still 2015, 28).

Garrett (2011), in particular, calls attention to the idea that more goes into making an effective, usable website than the interface we directly interact with on the surface. There are other multiple layers, all connected and influencing each other: at the bottom, the strategy level where site goals are defined; then the scope level where features and requirements are set; next, the structure level where behavior is governed; then the skeleton where information architecture and navigation are decided; and, finally, the surface where all the content, including language and images, come together to communicate to the user (Figure 2.8). Even decisions made at the strategy level can disrupt negatively what finally appears on the surface. Good UX must consider all aspects of the situation that is the production, and good UCD should find ways to involve user feedback and consider user needs and wants at every level, comprehensively. A user doesn't come to a website without experiences and expectations. But a website's surface isn't the only part that provides information about the site to bridge the gulf between expectation and evaluation and make the mental and conceptual models match up. The total user experience must be understood and designed with users involved.

It may appear that developing UCD has followed a pretty haphazard path. What this chapter attempts to show is that, as users became more involved using technology, designers tasked with making products for them have dedicated considerable resources to ensure what they make will be liked and used. Design thinking was originally focused on evaluating the capacity of the average human and then

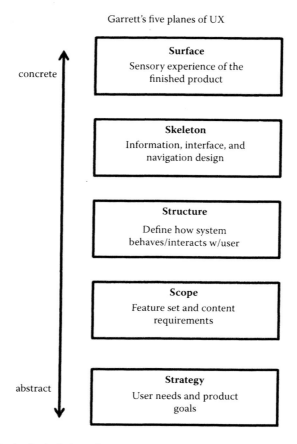

FIGURE 2.8   Author's depiction of Garrett's layers of the user's experience.

improving his or her performance—first to win wars while using advanced weap-onry and then to maintain the military industrial complex while using even more advanced systems. Until the PC revolution, users were a part of a system. The goal was a better system; therefore, humans were assessed and trained to participate ade-quately within that system. As people had access to more technology that was theirs, and as the WWW allowed people to shop online, consumers' purchasing power changed how users were treated. Usability testing attempted to accommodate people by finding interface problems and fixing them so that the system would work better for use. We are now in an era where technology is integrated into people's lives and designing for the total user experience has become the most logical choice.

TAKEAWAYS

1. UCD incorporates techniques from multiple disciplines that have been his-torically employed to better understand the user experience: interaction design, cognitive science, ergonomics, and user research, among others.

2. UCD has roots in participatory design, which began as a labor movement for workforce democratization. This method advanced from a technology-focused to a user-focused design with which user needs were understood and integrated into the design process.
3. Changes in workforce safety measures and industrialization during the World Wars were strong impetuses to study user performance. The goal was to focus on how relatively unskilled people could both efficiently produce and effectively use once expert-operated products.
4. Consumer goods that represented new technologies for nonexperts required designers to think about how to integrate them into the home and convince users that they could actually use them.
5. The arrival of the PC as a consumer product drastically altered the way developers approached the design process. With consumer satisfaction driving the market, designers had to focus on user needs and design a PC that was as easy as possible to use.
6. The availability of mobile devices led to complex systems in which the web became intertwined with our everyday lives. The result was greater complexity for users, which meant designers had to account for this complexity of the user experience.
7. To make products easy to use, usability testing became a popular methodology to involve user feedback into the design process. Representative users were asked to try out a product, and their performances were recorded and analyzed. However, usability testing was limited because it generally waited until the end of product development to test users or did not consider other factors that might impact the user experience.
8. An iterative, UCD approach that focuses on the user's mental model, environment, and complex user experience provides a broader, more accurate, situated, and dynamic picture of the user experience than usability testing alone.
9. The thoughts, actions, and needs of the user must be at the very center of the design process. Designers must consider the user experience and engage users as much as possible from the very beginning of the design process.
10. UCD incorporates users early and often and adjusts its understanding of users as well as the construction of the product being designed for them, based on the knowledge gained from interactions with and learning from users. Good UCD considers all aspects of the user experience at all stages in the design process.

DISCUSSION QUESTIONS

1. What are some historically expert technologies (e.g., pumping gas, driving a car) that you now use on a daily basis?
2. What are some current expert technologies that you think will be available to nonexpert users in the future?
3. How much should designers listen to users? One could argue that everyone "has" to use a computer in today's digitally connected environment, so why focus on what users want if they have to use it anyway?

4. What contemporary tasks would be difficult to complete without mobile computing technologies? How has the way you complete certain tasks or engaged with technology changed since the onset of mobile computing technologies (if you were around before then)?

5. Are there any parts of your life that are still "off-line?" Do you think these will become "online" in the future?

REFERENCES

Aguado, J. *Emerging Perspectives on the Mobile Content Evolution.* Hershey: IGI Global, 2015.

Albers, M. *Communication of Complex Information: User Goals and Information Needs for Dynamic Web Information.* Mahwah: Lawrence Erlbaum, 2004.

Aldrich, M. "History of Workplace Safety in the United States, 1880–1970." *Economic History Association.* 2001, accessed February 18, 2016. https://eh.net/encyclopedia/history-of -workplace-safety-in-the-united-states-1880-1970-2/.

Baddeley, A., and G. Hitch. "Working Memory." In *The Psychology of Learning and Motivation*, Vol. 8, edited by G. Bower, 47–89. New York: Academic Press, 1974.

Baek, E., K. Cagiltay, E. Boling, and T. Frick. "User-Centered Design and Development." In *Handbook of Research on Educational Communications and Technology*, edited by M. Spector, D. Merrill, J. Van Merrienboer, and M. Driscoll, 659–670. New York: Lawrence Erlbaum Associates, 2008.

Banbury, S., W. Macken, S. Tremblay, and D. Jones. "Auditory Distraction and Short-Term Memory: Phenomena and Practical Implications." *Human Factors* 43, no. 1 (2001):12–29.

Barnum, C. *Usability Testing and Research.* New York: Longman, 2002.

Becker, L. "90% of All Usability Testing Is Useless." *Adaptive Path.* June 16, 2004, accessed February 19, 2016. http://adaptivepath.org/ideas/e000328/.

Bias, R., and D. Mayhew. *Cost-Justifying Usability.* Boston: Academic Press, 1994.

Broadbent, D. "Donald E. Broadbent." In *A History of Psychology in Autobiography*, Vol. VII, edited by G. Lindsey, 39–73. San Francisco: W.H. Freeman, 1980.

DeGusta, M. "Are Smart Phones Spreading Faster than Any Technology in Human History?" *MIT Technology Review.* May 11, 2012, accessed February 19, 2016. http://www .technologyreview.com/news/427787/are-smart-phones-spreading-faster-than-any -technology-in-human-history/.

Ergoweb. "History." Accessed February 18, 2016. https://ergoweb.com/knowledge/ergonomics -101/history/.

Ericsson, K., and H. Simon. "Verbal Reports as Data." *Psychological Review* 87, no. 3 (1980): 215–251.

Foucault, M. *The History of Sexuality*, Vol. 1. Translated by R. Hurley. New York: Pantheon Books, 1978.

Garrett, J. *The Elements of User Experience: User-Centered Design for the Web and Beyond.* Berkeley: New Riders Publishing, 2011.

Genov, A. "Iterative Usability Testing as Continuous Feedback: A Control Systems Perspective." *Journal of Usability Studies* 1, no. 1 (2005): 18–27.

Gray, M. "Web Growth Summary." *MIT People, Matthew Gray.* 1996, accessed February 19, 2016. http://www.mit.edu/people/mkgray/net/web-growth-summary.html.

Halsey, A. III. "Eyes Down, Minds Elsewhere, 'Deadwalkers' are Among Us." *The Washington Post* (Washington: Washington D.C.), September 27, 2015.

Howard, T. "How Did Usability Testing Become a Dirty Word?" Keynote address at the Texas Tech University Technical Communication and Rhetoric May Seminar, Lubbock, Texas, May 22–June 3, 2015.

Hyman, I., S. Boss, B. Wise, K. McKenzie, and J. Caggiano. "Did You See the Unicycling Clown? Inattentional Blindness While Walking and Talking on a Cell Phone." *Applied Cognitive Psychology* 24, no. 5 (2010): 597–607.

Karpinos, B. "Fitness of American Youth for Military Service." *Milbank Memorial Fund Quarterly* 38, (1960): 213–247.

Kuhn, T. *The Structure of Scientific Revolutions.* Chicago: Chicago University Press, 1970.

Lavietes, S. "Alphonse Chapanis Dies at 85; Was a Founder of Ergonomics." *The New York Times* (New York). Oct. 15, 2002.

Miller, G. "The Magical Number Seven Plus or Minus Two: Some Limits on Our Capacity for Processing Information." *Psychological Review* 63, (1956): 81–97.

Molich, R., M. Ede, K. Kaasgaard, and B. Karyukin. "Comparative Usability Evaluation." *Behaviour and Information Technology* 23, no. 1 (2004): 65–74.

Morgan, J. "The Effect of Sound Distraction Upon Memory." *The American Journal of Psychology* 28, (1917): 191–208.

Nelson II, J. "General George C. Marshall: Strategic Leadership and the Challenges of Reconstituting the Army, 1939–41." *U.S. Army War College.* February 1993, accessed February 18, 2016. http://www.strategicstudiesinstitute.army.mil/pubs/summary.cfm?q=358.

Nielsen, J. "Applying Discount Usability Engineering." *IEEE Software* 12, no. 1 (1993): 98–100.

———. "Mobile Usability: First Findings." Nielsen Norman Group. July 20, 2009, accessed February 18, 2016. https://www.nngroup.com/articles/mobile-usability-first-findings/.

Price, B. "Frank and Lillian Gilbreth and the Manufacture and Marketing of Motion Study, 1908–1924." *Business and Economic History* 2, no. 18 (1989): 88–98.

Redish, J. "Expanding Usability Testing to Evaluate Complex Systems." *Journal of Usability Studies* 2, no. 3 (2007): 102–111.

Ritter, F., G. Baxter, and E. Churchill. *Foundations for Designing User-Centered Systems: What System Designers Need to Know About People.* London: Springer-Verlag, 2014.

Short, J. "How Much Media? 2015 Report on American Consumers." *Institute for Communication Technology Management, University of Southern California Marshall School of Business.* 2015, accessed February 19, 2016. http://www.marshall.usc.edu/faculty/centers/ctm/research/how-much-media.

Still, B. "Mapping Usability: An Ecological Framework for Analyzing User Experience." In *Usability of Complex Information Systems: Evaluation of User Interaction,* edited by M. Albers and B. Still, 89–108. Boca Raton: CRC Press, Taylor & Francis, 2011.

———. "Constructing a Better User Experience: Site Visits and User Shadowing." *Intercom* 62, no. 8 (2015): 28.

Telier, A. *Design Things.* Cambridge: MIT Press, 2011.

UsabilityNet. "ISO 9241-1: Guidance on Usability." UsabilityNet, 1998, accessed February 19, 2016. http://www.usabilitynet.org/tools/r_international.htm#9241-11.

U.S. Department of Education. "120 Years of American Education: A Statistical Portrait." *National Center for Education Statistics.* January 19, 1993, accessed February 18, 2016. https://nces.ed.gov/pubsearch/pubsinfo.asp?pubid=93442.

Valola, J. "Rough History of User Centered Design Disciplines." SlideShare. October 8, 2009, accessed February 18, 2016. http://www.slideshare.net/jonnevalola/rough-history-of-user-centered-design-disciplines.

Virzi, R. "Refining the Test Phase of Usability Evaluation: How Many Subjects Is Enough?" *Human Factors* 34, no. 4 (1992): 457–468.

Waterson, P. "World War II and Other Historical Influences on the Formation of the Ergonomics Research Society." *Ergonomics* 54, no. 12 (2011): 1111–1129.

Whiteside, J., J. Bennett, and K. Holtzblatt. "Usability Engineering: Our Experience and Evolution." In *Handbook of Human Computer Interaction,* edited by M. Helander, 791–817. New York: North Holland, 1998.

Williams, G., M. Welch, and P. Avis. 1985. "A Microcomputing Timeline." Byte, September (1985): 198–208.

Wolfe, T. *The Right Stuff.* New York: Farrar, Straus and Giroux, 1979.

Woodson, W., and D. Conover. *Human Engineering Guide for Equipment Designers.* Los Angeles: University of California Press, 1964.

Wozniak, S. "Homebrew and How the Apple Came to Be." *Atari Archives.* n.d., accessed February 19, 2016. http://www.atariarchives.org/deli/homebrew_and_how_the_apple.php.

3 UCD Principles

The word *methodology* describes a theoretical approach to the practice of something, complete with its own set of methods and principles that serve to execute the methodology's theoretical concepts. A philosophy, on the other hand, is the study of a particular set of ideas about a particular knowledge or a particular way of doing something. You may, for example, have a personal philosophy about how you should design for users and about what you should follow to ensure you are practicing correctly what you philosophically believe is the best way to design for a user. "Users are always right" or "you are never your user, don't design for you, design for users" are examples of philosophical mandates that might guide your UCD philosophy. But you might also have, for example, a prescribed process for how to carry out effective UCD, and should you also implement methods or declare and enforce in your practice of UCD certain governing principles, it could be argued that you are adhering to a UCD methodology.

Norman (2013), among others, has called human-centered design (HCD), or what we call UCD, a philosophy, not a methodology. Others have suggested that the focus in design of putting user needs and wants first is more a framework or way of thinking rather than a prescribed process with its own set of guiding principles to be followed consistently to guarantee effective results. Because almost every UCD situation is arguably unique given that it starts with a focus on the unique issues confronting a unique population of users using a unique product, to advocate for the same methodological approach to be applied to such a unique situation, and to others as well, might prove unwieldy if not ineffective. UCD, so the argument goes, is therefore more about establishing and reinforcing the belief (philosophy) that the user's needs and wants must come first.

But our problem with UCD conveyed as just a philosophical construct is that in the real world or the environment in which real design problems have to be addressed quickly and effectively involving various conflicts (e.g., budgetary constraints, timeline constraints, team personnel concerns, user knowledge limitations), a philosophy of UCD isn't enough. Practitioners, including designers, developers, project managers, and any other positions tasked with implementing a design project need to have a reliable set of principles to guide what they do, to measure success along with fundamental methods housed inside a reliable, repeatable process that they can carry out and adapt as necessary to involve users as the focus of their design.

Yes, UCD borrows from many other disciplines, as Chapter 2's overview of the historical underpinnings of UCD show, making it hard to assert that it has its own methodological construct. At the same time, there are genuine concerns about what the principles should be of a UCD methodology. As for process, or a way of implementing it from start to finish, we hear the voices of Norman and others claiming design, by its very nature, is too iterative, too contextualized, and too focused on finding the best solution, whatever it takes, to be able to adhere to a standard process of implementation.

But something, frankly, is better than nothing. We must be able to render to practice what we believe, and it must be a process, even with limitations, even one that may be modifiable, that can be carried out and integrated into already existing processes. This is all the more important for those organizations that have never carried out UCD of any kind. They may already have other guiding philosophies for how they define what they believe.

So often, startup software companies are heralded as great examples of how to carry out UCD. Of course, they often never had any previous approach in place for design. They started with UCD on Day 1 and don't know any other way to design. But for companies at which an agile development philosophy may already be in place, which emphasizes faster, better coding rather than UCD, or for those companies that may to this point have only done waterfall testing involving users at the very end of a product's development, UCD as a philosophy just isn't enough. It has to be there, but the way to introduce it into most organizations, and to reinforce it so that its tenets are spread and shared throughout, is to put in place an understandable methodological system, complete with core principles and methods for UCD. We tell you then that UCD must be both philosophy and methodology—something to be believed but also practiced.

In this chapter, we present what we consider to be the core principles of a UCD methodology. At its conclusion, we also discuss a process for implementing UCD that is consistent enough to be repeated and open enough to be modified to fit an organization's infrastructure or the peculiar demands of one product's design challenges compared to another's.

3.1   THE 10 UCD COMMANDMENTS (PRINCIPLES)

What follows are our 10 commandments for UCD:

    I. Thou must involve users early.
    II. Thou must involve users often.
    III. Thou must design for use in context (the product will be used in the real world, so design accordingly).
    IV. Thou must keep it simple.
    V. Thou must be polite.
    VI. Thou must know your users.
    VII. Thou must give users control.
    VIII. Thou must remember and design for emotion—people feel as much as they think.
    IX. Thou must trust but verify—triangulation is the key.
    X. Thou must discover before designing and delivering, and thou must know that discovery never ends, even after delivery.

If you adhere to these 10 principles, if they are achieved in the product or process you design to meet the needs and wants of the user, you can feel confident that you are taking the steps necessary to create effective UCD.

Before we look closely at each one, here are a couple of caveats:

1. This list, our 10 commandments, is by no means exhaustive. We accept the idea that there may be other UCD principles that can and should be considered. What we've done here, through our own experiences and research (Still 2016), and through research of the work by thought leaders in UCD, is find 10 bedrock principles that are often mentioned as being necessary to apply to construct worthwhile UCD experiences. For each one, we provide an explanation as well as examples to illustrate how it works and how to enact it.

2. In most instances, there is no rubric or scoring sheet you can apply when following these 10 UCD commandments. One cannot say that a particular UCD effort to build a project scored overall, for example, 82 out of 100 overall, or nine out of 10 for one specific principle. Certainly such quantitative measuring could be established. Other heuristics, including Nielsen's usability heuristics, offer the means to score in such a way. We encourage those who use our 10 UCD principles to think creatively about how they might be more quantified as actual numerical benchmarks for determining success in the real world, in the classroom, etc. But at this point, our purpose is to assert that they are qualitative guidance, rules that must be followed in one way or another to create effective UCD.

3.2   THOU MUST INVOLVE USERS EARLY

When is it too early to involve users? Never. And yes, we really mean that. How can it ever be too early to involve those people who will ultimately use your product? There are different ways that you can bring users in on the process, which we discuss in subsequent chapters, but to do UCD means that users and their needs and wants are the focus of the whole process. You might argue that user needs and wants can be considered later on, for example, when a product is rendered to some usable form, such as a prototype. But to do this risks building a prototype that already isn't right for its intended users, simply because you didn't from the very beginning consider your users. We point you to EyeGuide Assist (discussed in Chapter 1). No real effort was made to understand user needs and wants, and the result was a failed product.

Participation, as we noted before, can be weak or strong. It bears repeating again that to do UCD does not mean users are designers. But even if you don't have a means or desire to let users put their hands on the tools of design from the outset, you can still construct opportunities to involve them, and to do so will not only guarantee you're doing proper UCD, but it will also likely ensure you avoid the problems that often occur at the end stage of development when you discover certain design choices you've made don't work. Correcting such problems is expensive and time-consuming, especially compared to what it would have cost you in terms of time and money had you involved users at the beginning to learn what they needed and wanted.

Eric Raymond, one of the founders of the open source movement in software, wrote in *The Cathedral and Bazaar* (1999) about a primary difference between typical software development and release and open source software. In the typical process, just a few developers, working in isolation, create the code and release it to the masses for consumption; only those developers get to decide when it's ready, and they rarely take feedback during development, release, or even afterward. It is a cathedral approach. With open source, the central tenet is more eyeballs, fewer bugs; in other words, the more people working on software's development and release means more problems will be caught and fixed. In a bazaar where users and developers intermingle together creating code, then use it and give feedback, the larger community works more democratically to make more reliable software.

In this way, too, UCD's first commandment to involve users early functions metaphorically, much like Raymond's (1999) bazaar. As Still (2016) notes, "by focusing as soon as possible on user needs and wants, the design is exposed to more eyeballs, the important eyeballs of the users, and potential big problems are discovered and addressed before they become too big to be fixed" (26). As we have already pointed out, users have mental models of the world that allow them to navigate it. For products to be successful, they must be made for users to be successful and to integrate into their mental models, or they have to be constructed so that users with not a lot of effort, drawing on previous understanding, can adapt their mental models to use it.

3.3   THOU MUST INVOLVE USERS OFTEN

At eBay (Morgan and Borns 2004), user feedback is integrated throughout an ongoing process of redesigning the main eBay website. There is no end point to gathering this feedback. Users from around the world are recruited and given some money, and their experiences using the eBay website are then recorded and analyzed. The idea is simple: Constant user feedback means that eBay is always aware of what users think about its website. User reactions to new features can be tested quickly, modified, and reintegrated. Because the eBay website is the company's flagship, it cannot go down. Its look and feel also cannot be dramatically overhauled at the risk of confusing or alienating existing customers who have become accustomed to its display, navigation, and interactivity. To understand, therefore, how to tweak its website so that it performs effectively for users without downtime or too much of a drastic facelift, eBay integrates user feedback as often as possible by collecting, analyzing, and generating improvement as necessary.

UCD must involve users often. At the beginning of development, during rollout, and even after a product has gone live, users should be included. Surveys, focus groups, interviews, prototyping, user persona development, analytics, usability testing, journey maps, storyboards—we document and provide examples of the many ways to do it as we go forward (Still 2016). But users, to carry out UCD, must be involved. Iteration is core to UCD. We learn conceptually and early on how a user thinks about a product and what a user needs that product to do. As we create prototypes of the product, low in fidelity initially and then taking on a more refined form later, we have users use them.

In the process, we learn if content, the structure, the navigation, the interactivity, and the interface's look are all there and correct, if all of the layers of the user experience, which Garrett (2011) notes, have been considered and designed to be appealing and usable. And after the product has gone live and is in actual users' hands, we still learn from users by involving them. This is why Facebook conducts ongoing A/B testing; it gives some of its users a new feature, a B version, to use and learns whether that user group likes the B version better than the existing A version (Grant and Zhang 2014).

UCD designers must build into their processes ways for users to be integrated constructively. There is a challenge to do this in certain organizations that have set practices in place, such as agile development, which emphasizes moving quickly. But the Lean UX approach, growing in popularity, offers a solution that we examine in depth in Chapter 7 called the minimum viable product (MVP), which allows designers to quickly design and get feedback from users so as not to delay progress and push back delivery timelines.

Ultimately, users' needs and wants change, as do the environments in which products are used. Involving users often is both sound UCD and sound business. The data collected not only informs the development of the product in question, but it also adds to knowledge about users that can be used for other products.

3.4 THOU MUST DESIGN FOR USE IN CONTEXT

People don't use anything you create for them in a vacuum. Whatever product you design should accommodate the environment in which the user will work with it. Jeff Hawkins, inventor of the Palm Pilot mobile device, wanted to know how users would want to use what, at that time, the early to mid-1990s, was a relatively groundbreaking, mobile computing product. Users didn't have much of a mental model in place for using a Palm Pilot, and Hawkins was at a loss for what the Palm Pilot needed to satisfy users' needs in their daily contexts. He had a real business question, and rather than guess the answer, he literally cut a piece of wood (Butter and Pogue 2002) in roughly the shape he envisioned for the Palm Pilot (Figure 3.1). He created a stylus out of a chopstick and began to experiment with how people would use his low-fidelity "wooden" Palm Pilot in actual use scenarios, like checking email on the go or pulling up documents like spreadsheets on the commuter train. When Palm Pilot debuted, it was regarded as a highly usable, innovative solution for millions of business users.

Brian conducted user testing for a medical research publisher a few years ago. The publisher was working on a new electronic reference tool for physicians to use to consult for drug amounts, contraindications, and other information when treating a patient. The tool, popular to this point as a desktop application, was being adapted to a mobile environment, namely for iPhones and iPads. What Brian learned quickly through testing was that physicians had enormous trouble typing out symptoms, drug names, and the names of diagnoses because there was no auto-complete and no speech-to-text functionality. The system itself was, in fact, not an application created to work on mobile devices. During testing, physicians tried dutifully to complete the tasks assigned to them, but they were often frustrated. Some stated they had "fat finger disease," something that often plagues many people who type on mobile devices with keyboards much smaller than on a desktop or laptop.

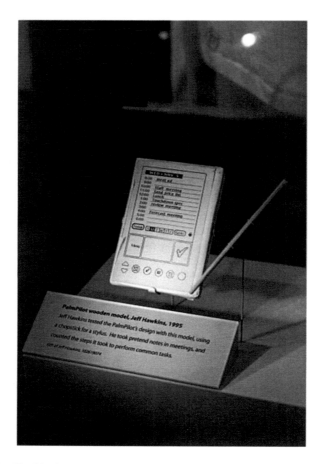

FIGURE 3.1 Hawkins' wooden model for the Palm Pilot. ("Wood/paper model for the PalmPilot, to try out screen designs and characterize the difficulty of performing tasks," Michael Hicks, November 3, 2013. Creative Commons Attribution 2.0, https://creativecommons .org/licenses/by/2.0/deed.en.)

Testing results like this spurred the publisher to ultimately create a mobile-friendly version, one that didn't require as much typing, had voice search, and even gave users a history or quick reference list they could access for their more common queries. UCD is about understanding and designing for people in the real world in which they live. We need to help our intended users mediate what is an often dynamic environment. If you know users will be doing other things while they work with the application you're designing, make it easier for them to do data entry rather than just type. If audio commands can be used to signal meaning that can support comprehension for the police officer busy viewing the road while driving, consider adding those in as necessary.

What Norman (2013) calls an *affordance*, a term we will use again, is the sound in the officer's car or the speech-to-text search in the mobile device. According to Norman, an affordance is "a relationship between the properties of an object and the capabilities of the agent that determine just how the object could possibly be used" (11). It isn't just the object, such as the mobile device, that has the affordance built into it. Rather, the affordance is how the user's capability is connected to the product's capability. If the user can use speech-to-text within the context of use, and if the product allows for voice-to-text search, we can call this relationship now established an affordance. Where will the product be used? What about the environment (e.g., indoors/outdoors, noise/silence, movement/stationary) can influence what the user needs and thus what the designer has to consider when creating capabilities in the product to meet up with the capabilities of the user?

A simple environment should never be assumed. Thou must design for the context of use.

3.5   THOU MUST KEEP IT SIMPLE

Simplify, simplify, simplify. Why is simplicity so important? We experience significant satisfaction and frustration each day, depending on how well or how poorly this principle has been applied to the things we use. For instance, think about a pair of scissors. You pick it up, you stick your thumb and fingers in the opposing loops, and you cut something. Simple. Now think about the last time you tried to reprogram the clock on your microwave. Of course, you probably didn't reprogram the clock because it proved too big of a hassle, and you gave up. And then maybe you wondered, "Why do I need a clock on my microwave anyway?"

Here are some more examples. Remember the first time you tried to use Adobe Photoshop to edit a photo? Now compare that experience with quickly cropping and posting a photo to Instagram. When using Photoshop, you are exposed to a slew of features, functions, and options, and it takes extensive training and tutoring to figure out how to use all of these features in concert with one another. Photoshop can do amazing things and gives the savvy user the capacity to perform artistic and specialized work, but it is not simple and easy to use. How many of the available functions are used on a regular basis and by what percentage of users? Most consumers may not want to use all of these features. Perhaps they just consistently want to change the contrast a bit and apply a red-eye filter.

Here's another example of how simplicity wins the day (and billions of dollars). You can purchase and download a book on Kindle with a brief search and a "1-Click" button, *or* you can buy it at a local bookstore but only after you find the keys, get in the car, navigate traffic, wander the aisles, talk to an associate to find the book, wait for him to look up the location of the book, go get the book, go to the check-out line, wait in line, hand over the book, swipe your card, sign the screen, take the bag with your item in it, and then drive home.

Is it any wonder why Amazon.com is so popular and why it has made it very difficult for traditional bookstores to compete? Amazon didn't just design a better app or website. It simplified the entire purchasing experience from the moment the customer makes the conscious decision to buy a book.

Complexity can be one of the greatest enemies of usability and one of the quickest ways to complicate a use scenario is to add a multitude of functions or features and then organize or reveal those features poorly. The opposite is also true. Fewer simultaneously available functions and well-organized features lead to simplicity and, essentially, ease-of-use.

Of course, the product must have some level of functionality to be considered valuable at all. The trick is that many products and designs are multifunctional, and smartphones are the shining example. The phones allow you to call people, text, send email, search the Internet, and download hundreds of apps that add hundreds of additional functions to the phone. Clearly, users like to have a convenient "one-stop shop" on hand. But the interface must ensure that it doesn't draw the user's attention to too many functions at the same time, or the user will experience cognitive overload and thus have a hard time performing tasks.

The right function or tool needs to appear just in time for the task or when the user seeks it. The challenge is balancing functionality with usability and predicting exactly which groups of users need/want specific functions, when they want to access them, and how. Giles Colborne's book, *Simple and Usable* (2011), championing simple design, breaks down users into the following groups: experts, willing adopters, mainstreamers.

3.5.1 EXPERTS

One key problem Colborne (2011) points out is that designers and engineers tend to develop products for experts like themselves—people who have specialized knowledge and grand expectations and also want a plethora of functions and possibilities, like a watch that doubles as a laser pointer or a personalized rocket ship. Experts tend not to mind complexity because it offers them a difficult problem to solve, and many experts are engaged in trying to solve unique and specialized problems in their fields. Experts represent the smallest population of users even though they tend to be the most vocal and capable problem solvers.

3.5.2 WILLING ADOPTERS

The willing or early adopter is someone who likes trying new things or is willing to take risks and explore the potential of what a product might do for them even if it doesn't work efficiently yet is expensive to operate or requires regular repair (Colborne 2011). Although early adopters can become trendsetters, they tend to have closets or app menus full of obsolete tools that turned out to be more trouble than they were worth.

3.5.3 MAINSTREAMERS

Mainstreamers, on the other hand, just need the thing to work and to save them some time and energy. They have little patience for troubleshooting. I buy the drill, I take it out of the package, I throw the instructions in the trash, and I use it to hang a shelf. Simple. Mainstreamers represent the typical user (Colborne 2011). They are practical and easily pleased or annoyed by a product's level of usability, familiarity, and emotional effect. The product needs to make them feel like they have more control. If the product is too complicated to use, too ugly, too foreign of a concept, too inconsistent, or too unreliable or untrustworthy, they simply won't use it.

UCD designers typically want to focus on meeting mainstreamers' needs and expectations even if those users don't really know what they want or need yet. Interestingly enough, if the mainstreamer likes to use a product, the experts and early adopters tend to like using them too. Although tolerance of complexity varies among these different groups, they universally tend to appreciate simplified and usable tools.

So how can we make something simple? Colborne (2011) offers a number of strategies that include prioritizing, removing, relocating, and hiding.

3.5.4 PRIORITIZE

What features are required to accomplish the primary function of the tool, process, or environment? A few years ago, as newer, "smarter" phones first emerged, some designs didn't demonstrate this priority. For example, a touchscreen feature on one model might "work" and allow you to navigate a limited number of helpful apps, but every time you put the phone up to your ear, you accidentally activate the touchscreen and drop the calls. In this case, the phone could no longer be relied on to accomplish its primary function. In another example, as flatscreens started to come out, some models offered beautiful graphic displays but the user could have a hard time finding the on/off button, an imperative and primary function.

A UCD designer defines the features that are absolutely necessary for the user to accomplish primary, frequently repeated tasks. Discover the features that do or don't contribute to primary tasks/purposes. From there, define the next most important features for conducting typical tasks and then the features that do not need to be there, no matter how "neat" they are. Discover what features your users need or want to access the most and work from there.

3.5.5 REMOVE

Another critical step is to remove unnecessary features. In some cases, features are superfluous, meaning the user doesn't need or want them. In other cases, the features shouldn't be included in the first place and belong in or should be used with a more appropriate tool or system. One good rule of thumb is to remove functions only the engineer/designer would know how to use.

3.5.6 RELOCATE

Another great way to simplify an interface, tool, or use environment is to move features in a way that puts the most heavily used features in the most immediately accessible place. For example, Brian's tablet interface used to have a screen capture feature placed in the primary menu, down at the bottom with the "home" button. He ended up taking plenty of screenshots but not intentionally. The most recent update to his tablet, fortunately, displaced this feature, moving it from the home menu to a "tools" app, which he needs to access only on occasion.

3.5.7   Hɪᴅᴇ

One of the best ways to help users focus on a task without being overloaded by other features is to temporarily hide other features or relocate these features in menus that the user can explore if they're interested. For example, when using the phone feature on Brian's smartphone, all other menu items or features unrelated to the call are hidden. Another way to hide features is to create defaults that work fine for the majority of mainstream users. Other nuanced features are hidden unless Brian specifically searches for them. If some users want to customize their features or tools, they can access them through menus.

   Best practices include hiding settings or options that you need to deal with only once, hiding specialized options or controls, and not requiring the user to customize anything—the feature or tool should work immediately without customization. In addition, ideally, the system will automatically or intuitively recognize, just in time, when the user has reached a point at which they need a feature or additional customization. And once they've resolved the issue, the system should also recognize that they don't need that feature anymore and will hide it immediately. For example, if you receive a call while playing Angry Birds on your phone, the phone automatically assumes that incoming calls take precedence over the game and shows the tools that allow you to answer or hang up on the caller. Once you're done with the call, the phone automatically takes you back to your game of Angry Birds, hiding the phone feature.

3.6   THOU MUST BE POLITE

Although UCD should ensure that a product is simple to use, users on their own, in most cases, will do everything they can to make it feel simple even when it isn't. Again, they are adapting the product to meet their needs in their world. Instructions are a perfect example. Even though following them step by step will likely improve the process for completing the assembly of, for example, a new cabinet for which the instructions are designed, users more often than not will forego the instructions, take a look at a finished version of the cabinet, such as a photo of it on the box all the parts for it came in, and then attempt to put it together without instructions. More than one user has remarked, "It just seems easier to do it that way rather than follow the instructions."

   The problem with so many instructions, like so many products, is that they are not created to be polite. Designers, focused on making the product work more than satisfying user needs and wants, will often program affordances for themselves into a product. In other words, they make sure the product works and will create whatever solutions they need even if those solutions are difficult for the user to use. Have you ever tried to open a door with a strange handle that doesn't, by its design, indicate at first glance whether it should be pushed, pulled, or twisted to open? How often have you seen such doors with a sign next to the handle telling you what to do, like push or pull? That may seem convenient, but it is a classic example of a designer building a product to achieve its technological end regardless of user needs or wants. The sign is the equivalent of a rude driver giving a quick wave right after cutting in front of you on a busy highway. It is an excuse to act badly, but it certainly isn't polite. A door with no clear way to open except for a sign telling you to push isn't very polite either, nor is a sign in small letters telling you that the door (although doors are for entering) is in fact not for entering (Figure 3.2). Users don't see this small sign until they've already tried and failed to enter through the door. Not polite.

The way most people deal with cognitive friction, when faced with a lot of challenges or possible choices that appear unclear or change without clear reason, is to simplify. They are essentially ignoring bad behavior and just trying to focus on what they need to do to get by. Spend some time watching a user having problems with something that is difficult to understand. Often users will move their faces and thus eyes as close to the difficult task as possible. They suddenly didn't lose the ability to see as well. Rather, they are almost behaviorally reducing their peripheral vision and source(s) of distraction so that they might concentrate better on accomplishing the difficult task.

But good UCD shouldn't make users do this. Good UCD should be polite. Cooper (1999), in fact, details 14 distinct ways that will make software (or really anything made for people to use) polite:

1. Polite software is interested in me.
2. Polite software is deferential to me.
3. Polite software is forthcoming.
4. Polite software has common sense.
5. Polite software anticipates my needs.
6. Polite software is responsive.
7. Polite software is taciturn about its personal problems.
8. Polite software is well informed.
9. Polite software is perceptive.
10. Polite software is self-confident.

FIGURE 3.2 A door that is not an entrance. (Courtesy of Dr. Brian Still.)

11. Polite software stays focused.
12. Polite software is fudgable.
13. Polite software gives instant gratification.
14. Polite software is trustworthy. (63)

What Cooper is addressing with these tenets are the unique expectations—or mental models in the case of UCD—that human beings have when they interact with objects. In the case of software, for example, Cooper writes, "Humans have special instincts that tell them how to behave around other sentient beings, and as soon as any object exhibits sufficient cognitive friction, those instincts kick in and we react as though we were interacting with another sentient human being" (62). In short, politeness builds user morale and makes the user feel in control. As a UCD designer then, you must be polite.

3.7   THOU MUST KNOW YOUR USERS

GOOB. We love acronyms. More on that later, but a really good one is GOOB or get out of the building. You really don't know who your users are unless you interact with them, unless you research their wants and needs, unless you study carefully how they use your products or others like it, and unless you are constantly in the process of teaching yourself more and more about who is the amazing, dynamic user for whom you design. The most important part of UCD, hence the first letter in that acronym, is user. We design for and include in that design the user.

You are not your user. We have often been told by designers, many of them very good at what they do, that because of their own lived experiences they have the right wealth of knowledge to understand the needs and wants of their users. But we don't agree.

Certainly, some people are really good at channeling what does and does not motivate users. But there really is no such thing as a "user whisperer." Besides, even if you are that cool, why risk relying on that? Confirm to everyone you really are a user expert by getting feedback from users that correlate with what you already thought.

For everyone else, start from the assumption that because you are designing for users, the primary thing you must do is to understand the experiences of users. Microsoft and Intel, to name a couple of companies, employ cultural anthropologists (Baer 2014), among other user experts, to study how people from various cultures and parts of the world incorporate products into their lives. It never pays to be the last one to do something. In an effort to make products that fit politely and simply and that don't make the user think but also work as required, many organizations big and small are focusing to increase efforts on grasping the motivations, emotions, capabilities, and preferences of users.

Earlier we mentioned Colborne's (2011) concept of the mainstream user. Others as well have written extensively on user types and how to determine them, many of which we cover in detail in the next chapter on user research. There are, in fact, dozens of methods to find out about users. Vermeeren et al. (2010) categorized ninety-six methods, ranging from contextual inquiries (also called site visits) to storyboarding.

We want to use more than one such method to triangulate our findings, much like how a GPS system operates, using more than one coordinate to calibrate and find a correct position. The goal is to get an accurate picture of our users and design for that and not what we think or guess might be right.

3.8 THOU MUST GIVE USERS CONTROL

Colborne, Cooper, Norman, the list goes on, all advocate for users to have or feel as though they have control when interacting with a product. We know that users will throw away instructions, ignore features, and take other steps to make a difficult-to-use product more manageable for them to use. Or they just won't use it. Or they will use it and catastrophic events will occur: flight control gear with identical knobs, impossible-to-understand shutdown instructors for a nuclear reactor, nonexistent or confusing dosage directions for children's medicine. In fact, for the latter, a 2010 study of the *New England Journal of Medicine* (Yin et al. 2010) pointed out that roughly 50 of 200 different over-the-counter children's medicines didn't have measuring devices nor did they provide understandable instructions for proper dosages, causing parents in many cases to guess on dosages that, in some cases, led to illness or even death for their children.

Control does not have to be actual features that give users the ability to make changes. As Endsley and Jones (2004) noted in Chapter 1, control also can be given to users just by providing them better situation awareness. Honestly, one of the greatest reasons why Google took hold so quickly in the early 2000s to dominate a once crowded search engine market was because it made users feel like they were in control of how they navigated for content online. It is ironic that once you enter a search in Google, often just using basic key word searches, you get back actually thousands if not millions of results that should feel overwhelming. But Google organizes them in such a way that users believe that what they want is right at the top, which is why most users never go too far into their search results. Lee reports (2013) that the search-targeted advertising company Chitika found in a study of search results that the first search result returned was selected 33% of the time by users, the second result 18%, but the 10th result, not the thousandth but the 10th, was only selected a little more than 2% of the time. These findings indicate that people think they found what they wanted right away when searching with Google even though they just entered a basic key word or phrase search. Control, or the feeling of control, is essential to UCD as it is often to good business.

A few years ago, Brian was brought in by a hotel chain to evaluate the reasons why, despite a lot of money spent on purchasing and installing new radio/alarm clocks, customer phone calls to the front desk for wake-up calls had gone up, not down. Customer complaints were also more frequent than before the hotel chain had installed the new clocks. One of the first things Brian and his team did was study the clock and attempt to understand how a typical hotel guest, who was usually a business traveler staying for only a night or two, would want to use the clock. They were struck by just how sophisticated the clock was, including music that could be selected by genre and an alarm that could be set for wake up as well as for a snooze or a nap. Users, in other words, were given a lot of options, which designers sometimes translate as giving the users more control.

But control is not about unlimited functionality. In the early 2000s, many thought the portal design for websites was the answer to making users happy. Companies spent millions on creating complicated websites that gave users multiple features, based on their preferences, for visualizing or manipulating information. But users didn't want that. They wanted simplicity, and they wanted control. One need only look at Amazon's "1-Click" as a testament to this. "There's no way people will store their credit card information just so they can order something quickly online. They'll be too scared of someone hacking their account and buying whatever they can." That was the thought, but hundreds of millions of dollars later, we know now that Amazon 1-Click was transformational. It made buying easier. Yes, there were many choices, but users controlled that by searching for what they wanted. Suggestions were made for them based on previous purchases and other data, but none of that interfered with users being able to navigate quickly and purchase easily.

Control is about intuitiveness, in context. Does it make sense? Can you use the product to do what you need or want it to do in the situation? When you interface with it in this situation, does it impart a feeling that you can use or control it as you need or want? The problem with the hotel alarm clock was that it asked the user to do too much or gave so much for the user to do in a context in which the user, likely overwhelmed with a busy travel schedule and only staying for a night or two, just wanted a clock that could be controlled quickly and easily to set a reliable wake-up time. Because the clock didn't give users what they needed, they chose the simpler solution of calling the front desk and asking someone to wake them up.

We often say users are like water. Water will find a way to go downstream regardless of obstacles presented to it. It will flood over the top, go around or under barriers, and seep through cracks. Water knows where it wants to go and does everything it can to get there. We can try to stymie it, but to do so is expensive and never guaranteed. We build on high water tables and then have to install sump pumps in home basements or underground subway and train systems so that they will, working constantly, remove the water trying to flow in. Of course, we need basements and subways, and we have to build them somewhere. Keeping the water at bay makes sense. But such an argument falls flat when discussing users. Users are like water, but rather than attempt to block them from doing what they want to do, why not just design effectively to give them control or a feeling that they can flow where they would like to do what they want or need?

3.9 THOU MUST REMEMBER TO DESIGN FOR EMOTION

People feel as much as they think. They make up their minds in as little as four milliseconds (Zajonc 1980, 2000) often based on emotion more than logic. Phillips and Chaparro (2009) reported on a software usability study that indicated people perceived a website to be more usable based on the visual appeal of the site even if the site was not as usable as another website that had less visual appeal. As Norman (2002) writes, "Usable designs are not necessarily enjoyable to use" (8).

This is why, as Still (2016) notes, "we cannot design just for usability" (27). Certainly, effective UCD requires usability, but building something to be effective, efficient, and satisfying is not the same as building something that is also emotionally appealing. According to Norman (2002), the design of a product strikes us at three levels emotionally: visceral, behavioral, and reflective.

Visceral: This is the appearance of a product and our reaction to it—what is our gut reaction when we see it? Does it appeal to us? Does its design make us want to pick it up and use it?

Behavioral: This is our feelings about using the product—do we like using the product? Do we feel comfortable using it?

Reflective: This is how we see ourselves as we use the product—can we see ourselves using it? Do we like the idea of that image?

A well-designed UCD product satisfies our needs and wants, helps us work better and more efficiently, is one we like, one we feel comfortable using, or even one we like the idea of using or have no aversion to using. It is a product that is, therefore, designed with the assumption that emotion and cognition are inextricably intertwined. One cannot have one without the other and have an effective product that people want to use and, when using it, use well.

If you think about a product that meets all of these criteria for you, it is likely one of the favorite things you use. Norman (2002) writes eloquently of his favorite teapot. For Brian, it is his Colorado cabin's wood stove (Figure 3.3): a compact black base with a long black pipe extending elegantly into the wood-paneled ceiling. The glass door is easy to open with a strong handle and latch. The control for airflow smoothly moves from one setting to another. Once a fire is lit, not only are the orange and yellow glow of the wood appealing, but so is the crackling sound of the wood as it burns. The stove does not overwhelm the room it occupies, but it is the focal point of it. Sound, look, use, even smell—all of the senses are affected. At stressful times, Brian likes to imagine himself putting wood into the stove, starting a warm fire, and then enjoying the results. The stove appeals to him viscerally, behaviorally, and reflectively.

Obviously, software products, websites, and other digital technologies offer challenges that make them more difficult to appeal emotionally to users. This is especially true for any that might be used in the workplace where users are in contexts in which they have to use such products. Already, in some cases, emotionally positive feelings are compromised by the influences of the job. Still, even in the office, those products that users do better with or don't mind using are those that are designed for feeling and also thinking users.

Just recently while traveling for a conference, Brian ate at a very busy Indian restaurant in Logan, Utah, that was housed in a building that shared space with a convenience store and gas station. The manager of the restaurant also managed the convenience store! It was a busy Friday night, homecoming for Utah State University, so the manager was not only greeting and seating customers at the restaurant as well as taking phone orders and processing transactions; she was, at the same time, also walking through a short hallway and handling customers at the convenience store.

FIGURE 3.3   Wood stove. (Courtesy of Dr. Brian Still.)

The wait to get seated at the restaurant was fairly long, so Brian spent his time watching the manager go back and forth to deal with tasks at both businesses she supervised. He noticed she had a touchscreen monitor in both locations to manage transactions. This allowed her to enter necessary items, such as payment for chicken korma and naan at the restaurant or beer and chips at the convenience store, without having to do any sort of heavy manual work, such as typing. Each business had its own door, and a wall partitioned one from the other, but the registers were positioned closely to each other, and she was faced so that she could see traffic coming into either business.

Brian couldn't help but ask her what it was like to manage all of this work at the same time. He wondered how it made her feel, if she felt overwhelmed and hated doing it. She said no, that it was easier now than before when there were two different systems for handling transactions. When the owner got her identical touchscreens to manage activities at either place, she was able to work faster, even in busier times like that night, to get the job done. She also didn't mind how much was going on because it meant the businesses were doing well, and that was important to her. In this case, then, the technology, although not as enjoyable to use as something in a relaxing home environment, was still emotionally effective because the manager saw it as easier (visceral), enjoyed using it more than the previous solution (behavioral), and was proud (reflective) that it allowed her to work better to make the businesses successful.

If you don't consider how users feel about products, you risk building something effective, something arguably even usable, but not something that people enjoy using, which might mean they don't use it as well or at all. There are questions designers should always ask to verify that our products are emotionally appealing. There are even heuristics and more formal checklists to use, which we'll discuss later. Certainly, however, we must do what it takes to design for emotion. What people feel determines what they want to use and how well they use it.

3.10 THOU MUST TRUST, BUT VERIFY— TRIANGULATION IS THE KEY

Доверяй, но проверяй (doveryai, no proveryai), or trust but verify, is a Russian proverb. President Reagan was fond of using it to describe negotiations with Russian (then Soviet Union) leaders regarding nuclear arms treaties and other issues Reagan was willing to trust, but any such trust must be verified with additional sources of truth.

Such is also true for UCD. We should trust our instincts as designers. We should trust what we learn from talking to users or from carrying out other types of research that provide us with information about what users want or need. But we need to verify these instincts. We are forever in a state of verification: designing, getting feedback, testing, designing, and getting more feedback.

Triangulation is the key. It is a data analysis approach borrowed from qualitative research, which involves more descriptive research, based on observation, which is less capable of measurement than quantitative research (a hint to understand the difference is that *quali*tative is about *quali*ty of data, what we can see and make connections about, and *quant*itative data is what we can *quant*ify or measure about data, what the numbers mean when measured). Because qualitative data analysis is about answering why or how more than what, the answer isn't something easily configurable like $2 + 4 = 6$. The user's experience, such as it is, is also resistant to such formulaic conclusions. Why are users the way they are? How do they work? How can we make something that works for them to be effective and enjoyable? How do we know we have done this? Certainly, we can apply formulas to learn some things about users. Usability testing is highly necessary for confirming the success of a product, and it relies on a number of quantitative tests and other methods.

But if we want to understand the wants and needs of our users, if we want to grasp and design for those wants and needs within a dynamic, particular environment of use, we need to collect as much observable data as we can, throughout the design process, and then compare it to see what commonalities we can find. To do this, we triangulate. We don't rely on one source of information gathered from one method of user research. We employ at least three different methods, each one a little different in how it works, to gain knowledge about our users. And we do this at every stage of design, including after the product is live and being used.

For example, eBay, as Morgan and Borns inform us (2004), started with surveys, interviews, and focus groups to learn more about its users. But it didn't just trust that data alone. It looked at existing analytics based on real use of the eBay website. It then went further. Users were asked to carry out testing of prototypes of the eBay website, and data from that was collected and compared to the other user research data. Finally, even as the new changes went live, it continued to test users, collect analytic data, and also use surveys and interviews. All of these methods and data sets allowed eBay to create a more trusted and verified view of user needs and wants.

In our usability research, we preach the *see-say-do triangle*, which we talk more about in the next chapter. It's an easy way to remember that for every UCD situation, you should use what people say (verbalization), what you see them do (observation), and also what they do (performance) when you make design decisions focusing on their needs and wants. Users may tell you they want the design interface to be a certain way, but unless you verify that request with what they do with a test version of it, you don't really know if what they say they want is what they actually need to be successful.

We trust, but we verify.

3.11   THOU MUST DISCOVER BEFORE DESIGNING AND DELIVERING, AND THOU MUST KNOW THAT DISCOVERY NEVER ENDS, EVEN AFTER DELIVERY

Let's think of the complete UCD process, including the aforementioned commandments, as a universe of ideas, efforts, methods, goals, and requirements. It is an arrangement of multiple things all interconnected, all being changed and changing each other as the product being developed is first born, is first used, is modified based on feedback, and then matures into a final deliverable that is then still used. Knowledge from that engagement continues to shape it as it also influences new products in the same UCD universe.

A star metaphor works well here. In the UCD universe, the star or product is born from other events, spinning off from some dying supernova implosion, only to become the center of its own system, gathering planets with its gravity, shaping them and what they do as it has been shaped.

There's a lot going on in the UCD universe that contains the product you're making, and that process never ends. We are, in other words, constantly discovering, designing, and delivering. This is true for every product we make for users; in fact, we are discovering, designing, and delivering at every stage of that product, including the very beginning as well as the end.

To answer the question, "how do you do UCD?", the answer, and our final commandment, is that we discover before designing and delivering, and we never stop discovering.

Discover: Discovery is all the time because discovery is the process of understanding the following:

- Who our users are
- What their needs are
- What we need to do to make our product work for them
- What others have done
- What we can learn from users to improve this product and others

We never stop discovery as it is the most important thing we do to evolve one product, to make it usable, and to then spin off another usable product and then another.

Design: Design flows and feeds discovery, but we are constantly designing for users and getting feedback from users, and by iterating design, we improve through stages, in control but moving quickly; design doesn't and shouldn't be one perfect attempt but instead should start with anything, even rough things like paper prototypes, so that user feedback can be integrated, changes can be made, and the design can be enhanced.

Deliver: At some point, the thing we're designing has to be used to satisfy our goals (to make money, to inform), and the ultimate goal of UCD is to deliver an effective, usable product to its intended audience.

And yet the discovery process doesn't end at this delivery point. We learn from the delivery, from the initial reaction of the product to the ongoing use of it, and it feeds better UCD for new versions of that product or new creations of others.

UCD is about the ongoing process of discovering from users what is necessary to make the product usable for them, designing it to meet those needs, and delivering a product that satisfies those needs while also achieving your design goals and meeting your organization's requirements for success (Figure 3.4).

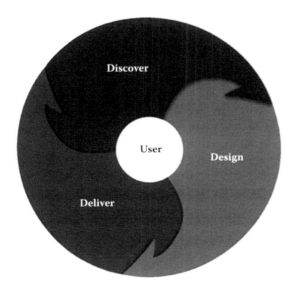

FIGURE 3.4   The discovery, design, and delivery process for effective UCD.

If these 10 UCD commandments offer mandates for what we must do to deliver UCD, how do we carry out UCD as a repeatable process, complete with methods and other practical tools? What is the most effective way to implement UCD and its guiding methodological principles?

3.12   RABBIT

We warned you we love acronyms. Mnemonic devices are great user-centered memory enhancement tools. Once people are taught certain core concepts, the mnemonic device is introduced. Remember that device, and studies show people can remember more easily what the device stands for. Acronyms, one type of mnemonic device, are designed so that each letter in the acronym represents something that identifies the object, concept, etc., to be remembered. Some acronyms are so effective at doing this, in fact, that they become part of our lexicon. SCUBA, which we now usually don't bother putting into all caps, is one of the more famous examples; it stands for self-contained underwater breathing apparatus.

We're not going to promise you that RABBIT will be as effective as SCUBA, but it is intended to help you remember the key steps that make up the UCD process. There are a lot of things involved in each step, and sometimes they may not be all used, or you may introduce your own, or you may even dare to ignore steps, if not mix and match their order. There are no hard and fast rules other than that you are centering the design process on the user who will end up using what you create.

But we like to think of the UCD process that should be carried out every time as *RABBIT*:

- *R*esearch users.
- *A*ssess the environment, the project goals and requirements, and the competition.
- *B*alance user needs and affordances with your team's design aesthetic and overall organization demands.
- *B*uild out an operative *I*mage for users to interact with that matches where you are in the project.
- *T*est the image with users to gather feedback.

For every UCD project, endeavor to follow these steps to ensure success. You don't necessarily have to follow them in order, and sometimes methods or techniques found in one of them can be repurposed in another. The visual in Figure 3.5 gives you a better sense of how all of the steps of the RABBIT process can flow together.

The important thing is that you make sure RABBIT is how you implement your UCD process because it guarantees you've included all key considerations necessary to make your product centered on user needs and wants.

RABBIT steps are complex, and we've devoted a chapter to each one with detailed explanations of useful methods and techniques as well as examples. To show how all the steps can be connected together in one successful process, we have also included a case study of a product, EyeGuide® Focus, built using RABBIT.

Focus served as EyeGuide's effort to involve users throughout the design process. Therefore, users were researched and given a say before any code was created, the

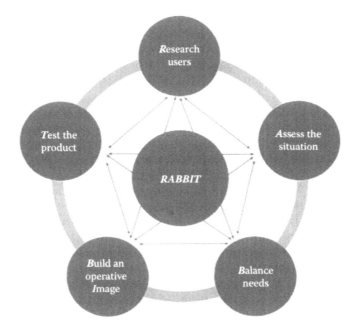

FIGURE 3.5 The *RABBIT* process for UCD.

situation or environment in which the product would be used as well as competitor offerings were assessed, the company's goals along with user needs and wants were balanced to generate a features set, and a series of prototypes were built and tested by users with results put back into the design—in other words, the RABBIT process was implemented.

TAKEAWAYS

1. UCD must be both philosophy and methodology—something to be believed but also practiced.
2. Focusing on user needs early in the design process helps address problems before they become too big, too expensive, or too time-consuming to fix.
3. Adapt the design to the user. Don't expect users to adapt to the design.
4. Successful products integrate a user's mental model into the design. A product should be constructed so that users (with not a lot of effort and drawing on previous understanding) can adapt their mental models to use it.
5. UCD is about understanding and designing for people in the real world in which they live. A simple environment should never be assumed.
6. Adding a multitude of functions or features or organizing or revealing features poorly quickly complicates a use scenario. Less simultaneously available and well-organized functions and features lead to simplicity and subsequent ease of use.
7. Focus on meeting mainstreamers' needs and expectations. If the mainstreamer likes to use a product, the experts and early adopters tend to

like using them too. Although tolerance of complexity varies among user groups, they universally tend to appreciate simplified and usable tools.

8. Define the features that are absolutely necessary for the user to accomplish primary, frequently repeated tasks.

9. Most people try to simplify when they encounter cognitive friction (being faced with a lot of challenges or possible choices that appear unclear).

10. You really don't know who your users are unless you interact with them, research their wants and needs, study carefully how they use your products or others like it, and are constantly in the process of teaching yourself more and more about the amazing, dynamic user for whom you design.

11. Use more than one UCD method to triangulate your findings. You should document what people say (verbalization), what you see them do (observation), and what they actually do (performance).

12. Users like to feel that they are "in control" when they interact with a product. Control does not have to be actual features that give users the ability to make changes; it can just mean more situational awareness. Control is about intuitiveness in context.

13. A product that satisfies our needs and wants, that helps us work better and more efficiently, that we like and feel comfortable using, or simply like the idea of using and have no aversion to using is a well-designed UCD product.

14. Discovery is the most important step in the UCD process.

15. If you don't consider how users feel about a product, you risk building something effective, something arguably even usable, but not something that people enjoy using. Design for emotion. What people feel determines what they want to use and how well they use it.

16. UCD is about the ongoing process discovering users and discovering what is necessary to make the product usable for them, designing it to meet those needs, and then delivering a product that satisfies those needs while also achieving your design goals and meeting your organization's requirements for success.

DISCUSSION QUESTIONS

1. How might you quantify one of the 10 UCD commandments as an actual numerical benchmark for determining success of a user experience in the real world?

2. Note the types of environmental influences typically present when you conduct an activity in your classroom or workplace (e.g., location, space, temperature, lighting, sound level). How is the context in which you conduct your activity impacted by environmental factors? Does the context make it easier or more difficult to complete the activity?

3. Many users are "intermediates/mainstreamers" who just learn the minimum number of features necessary when using a product to complete a specific goal. Given this, why do companies create so many product features in the first place? Is it better business to give users everything they could possibly want or just enough features to satisfy them?

4. What do you do when you encounter cognitive friction during a use scenario? How does it influence your thoughts, actions, or beliefs about the product?
5. What is a product that you love to use? How does it satisfy your needs and wants, help you work better, provide comfort while using, and help you to like the idea of using it or have no aversion to using it?

REFERENCES

Baer, D. "Here's Why Companies Are Desperate to Hire Anthropologists." *Business Insider.* March 27, 2014, accessed February 25, 2016. http://www.businessinsider.com/heres-why-companies-aredesperateto-hireanthropologists-2014-3.

Butter A., and D. Pogue. *Piloting Palm: The Inside Story of Palm, Handspring, and the Birth of the Billion-Dollar Handheld Industry.* New York: John Wiley & Sons, 2002.

Colborne, G. *Simple and Usable Web, Mobile and Interaction Design.* Berkeley: New Riders Publishing, 2011.

Cooper, A. "14 Principles of Polite Apps." *Visual Studio Magazine* June (1999): 62–66.

Endsley, M., and D. Jones. *Designing for Situation Awareness: An Approach to User-Centered Design.* New York: CRC Press, 2004.

Garrett, J. *The Elements of User Experience: User-Centered Design for the Web and Beyond.* Berkeley: New Riders Publishing, 2011.

Grant, A., and K. Zhang. "Airlock–Facebook's Mobile A/B Testing Framework." *Facebook Code.* January 9, 2014, accessed February 25, 2016. https://code.facebook.com/posts/520580318041111/airlock-facebook-s-mobile-a-b-testing-framework/.

Lee, J. "No. 1 Position in Google Gets 33% of Search Traffic." *Search Engine Watch.* June 20, 2013, accessed February 25, 2016. http://searchenginewatch.com/sew/study/2276184/no-1-position-in-google-gets-33-of-search-traffic-study#.

Morgan, M., and L. Borns. "360 Degrees of Usability." Proceedings of the Conference on Human Factors in Computing Systems, Vienna, Austria, April 24–29, 2004: 795–809.

Norman, D. *The Design of Everyday Things.* New York: Basic Books, 2013.

———. "Emotion and Design: Attractive Things Work Better." *Interactions Magazine* ix, no. 4 (2002): 36–42.

Phillips, C., and B. Chaparro. "Visual Appeal vs. Usability: Which One Influences User Perceptions of A Website More?" *Software Usability Research Lab.* October 15, 2009, accessed February 25, 2016. http://usabilitynews.org/visual-appeal-vs-usability-which-one-influences-user-perceptions-of-a-website-more/.

Raymond, E. *The Cathedral and the Bazaar.* Sebastopol: O'Reilly Media, 1999.

Still, B. "Fundamentals of User Centered Design." *Intercom* 63, no. 3 (2016): 26–27.

Vermeeren, A., E. Law, V. Roto, M. Obrist, J. Hoonhout, and K. Väänänen-Vainio-Mattila. 2010. "User Experience Evaluation Methods." Proceedings of the 6th Nordic Conference on Human-Computer Interaction: Extending Boundaries, Reykjavik, Iceland, October 16–20: 521–530.

Yin, S., M. Wolf, B. Dreyer, L. Sanders, and R. Parker. "Evaluation of Consistency in Dosing Directions and Measuring Devices for Pediatric Nonprescription Liquid Medications." *The Journal of the American Medical Association* 304, no. 23 (2010): 2595–2602.

Zajonc, R. "Feeling and Thinking: Closing the Debate over the Independence of Affect." In *Feeling and Thinking: The Role of Affect in Social Cognition,* edited by J. Forgas, 31–58. Cambridge: Cambridge University Press, 2000.

———. "Feeling and Thinking: Preferences Need No Inferences." *American Psychologist* 35, no. 2 (1980): 151–175.

4 Research Users

A few years ago, Brian and his lab team were contracted to help a university law school rebuild its website. In addition to three usability tests of site prototypes developed, Brian's team was also tasked with discovering the likely users of the website. The law school's aim was to make its new site a recruiting tool. The project manager told Brian during an early meeting, "Here are the three law schools in the region ranked ahead of us. We want to be better than them. We want our website to be better than their websites so that we can attract students that go there to come here."

It isn't unusual and, in fact, it makes a lot of sense, for organizations to think of their websites as flagships, the most forward-facing part of the organization that attempts to signal to the world, and especially to prospective customers or, in the case of the law school, students, that this is who we are and what we can give to you if you buy what we're selling.

At the outset, Brian asked the client (the law school), "Who do you see as your users? Do you have a goal for recruiting particular users since you have indicated you want to be competitive with other, presently more highly ranked, law schools?" The client indicated confidently that it knew its primary audience of prospective law school students was made up of, on average, younger than typical law school students, around ages 23–25. In addition, males more than females would be in this audience, and ethnically they were predominantly white. As the project manager noted, "We see students interested in our law school wanting to get their degree and settling down to practice or work in government around here," as in the law school's region of the country.

Had Brian taken this information from the law school as prima facie accurate, the final website designed for it would have looked quite different than what it eventually became. But we cannot ever rely on just one source of information to make determinations about users. We don't ignore the client. We don't ignore other data. But we always triangulate—trust but verify. Brian dutifully wrote down the law school's assumptions about its audience, but then he and his team set about discovering other sources of information about the prospective law school audience that would either confirm what the law school thought its audience was or would shed new light on possibly more accurate depictions of the actual audience.

Eventually, through due diligence including surveys and data of activity from the law school's present website, Brian and the lab team learned that prospective students of this university law school, and thus those the law school both needed to recruit for feedback during the new website's design and also design for, were not whom they assumed. Prospective students were actually around age 30, female, more ethnically diverse, and interested in taking advantage of an affordable law school education in this region and using it as a jump-off tool to move to a larger metropolitan area with more opportunities for private practice or government work.

Because the UCD process is about involving users as early as possible in the design process and placing your focus on them as you make something for them, it is important that you understand who will ultimately use the product you're designing. You want to know everything you can about the following:

- What users want and need
- What they experience in their use environments
- What motivates them to use certain products
- What makes it possible for them to use these effectively
- What obstacles stand in the way of successful use or problem solving

Had the law school built a website with feedback from the wrong audience, although its website may have eventually scored well in terms of usability, it would have likely excluded critical content appealing to the needs and wants of the actual audience.

You can learn about users only if you research them. You cannot make assumptions about who may be your users. You also cannot rely solely on in-house exercises, such as brainstorming, to determine your users' goals, needs, wants, and environmental influences. To understand your users, you must draw from a variety of methods, including those that involve direct interaction with users where they work and live. As Kujala and Kuappinen (2004) note, "At its best, user identification is based on both quantitative and qualitative information including market segmentation studies and field studies, etc." (6).

Before jumping into gathering data, you first need to determine what kind of data is appropriate for your design goals. Are you creating a new product? If so, researching how users interact with competing products or observing how they currently go about their tasks without your product may be the best way to start. Are you developing a new version of an existing product? Visiting user forums to see how users are discussing the product may provide insight that direct interviews would not. However you decide to begin your research, you want to select at least three methods: one that captures what users actually do, one that captures what users say, and one where the designer "sees" how users interface with design. We call this the *see-say-do triangle* (Figure 4.1).

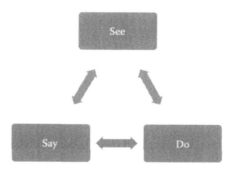

FIGURE 4.1   The see-say-do triangle.

Data analysis	Observation	Self-reporting	Designer analysis	User and use diagrams
Google analytics	Site visits	Focus groups	Task analysis	User flows
Site stats	Shadowing	Surveys	User matrix	User journey maps
Market research	Fly on the wall	Interviews	Use cases	Storyboard
User forums/reviews	Emotion assessment	Diary studies	Affinity diagramming	Card sorting

FIGURE 4.2   User research methods.

Think about how a GPS navigational system works. It doesn't rely on just one source of information to pinpoint a location. Rather, it triangulates multiple points, each one used to confirm the other. By triangulating the user research methods available to you, you can note patterns, errors, gaps, and even incidental actions from different sources, each one representing a different approach: user performance, user verbalization, and designer observation. Because these different methods provide different data from users' perspectives, their demonstration, and the designer's observation, you can have more confidence when you learn something about your users because it is confirmed by more than one source of data.

There are many methods you can use to understand who users are, what motivates them, how they work, and where they work. In turn, you can use this user data to begin designing a product that clients will find useful and desirable. As seen in Figure 4.2, methods are categorized into five different types: data analysis, observation, self-reporting, designer analysis, and user and use diagrams.

In no way do we expect or recommend you use all of these research methods. This list also isn't exhaustive. Look around online and you'll probably find many others. Choose methods that best suit the purpose of your design and user inquiry. Further, these methods can certainly overlap and fulfill more than one point on the *see-say-do triangle*. For example, if you conducted a site visit and used think-aloud protocol to collect data, this would count as both see and say. You can use any of these methods, merge methods, and reuse methods, but ultimately, the goal is to triangulate.

In the following sections, we go into more detail about each category and method, the benefits and drawbacks of each, and examples to put these methods into context. The end goal will be that you have the information you need to create profiles of users that allow you to recruit representatives to test prototypes of your product. Alternatively, you also will be able to create user personas, which are "fictional representations of people that are created from real data about your users" (Barnum 2011, 94). Either result, or a combination of both, will get you one step closer to building a UCD product. Remember, everything we do begins and ends with users.

4.1   EXISTING DATA ANALYSIS

There is good reason why existing data analysis is the first category of methods discussed. In all research, not just UCD, examining data that already exist not only helps you avoid recollecting data (which saves time), but it also allows you to delve deeper into your other methods by asking questions or looking for material that is new and important to understanding the context of your design. Therefore, finding published research or raw data from the organization or client you are working with should be your first step. Specifically, let's focus on Google Analytics and other statistical material, market research, user forums, user reviews, and internal computer support data.

4.1.1 Google Analytics and Statistical Data

This method of research requires delving into quantitative data that already exists or is auto-generated. Quantitative data can be counted (and is usually expressed in numbers and statistics). These types of resources provide broader, more in-depth views of the entire user population's activities. With that being said, we don't rely solely on this broad data because it does not give us a complete picture of users' motivations, nor can we accurately infer what specific obstacles may have caused issues with a product. However, this data can be useful when it comes to vetting the viability of particular products or features, especially if you've been encouraged or instructed to build products that key stakeholders want. Strong data sets from actual user experiences help us understand whether such products will be used or, if used, what features need to be included to make eventual adoption successful.

Google Analytics is one example of this kind of quantitative data. Essentially, Google Analytics captures the traffic flow of websites and illustrates who is viewing the material, users' locations, what pages they are visiting first (e.g., the "landing page"), and what devices are used to view websites. This information can tell us quite a bit about the users in question. For instance, if you are conducting user research for a new ecommerce website, you may want to use Google Analytics of an existing ecommerce website similar to your own. Doing so, we get an idea of who target users are, where they are located, what information they most often seek first, and what kind of devices they use to access the website. What does this tell us? First, we get invaluable demographic data. Second, we can begin to understand the ways users use the website (i.e., they look at specific products via their iPhone).

Take a look at Figure 4.3 to see an example of Google Analytics for Grinbath .com. Not only is the total number of visitors to the site calculated and presented, but further details, such as the number of page views and the average length of each user session, are also shown. Such data can give you a good sense of who is using your product, for how long, and even what pages or parts of the product they use. Often, you may have a hunch about what people like, and you may also ask them, and they will give you answers, but if for some reason what analytics actually show you is different, you need to understand why. We look at analytics like this to help us both gain a quantitative, big-picture understanding of how users use something and verify other types of data we collect from users.

Other statistical data sources can be, and should be, used as well. The Pew Research Center, for instance, reports research on many technology-related topics. In "U.S. Smartphone Use in 2015," Aaron Smith (2015) of the Pew Research Center

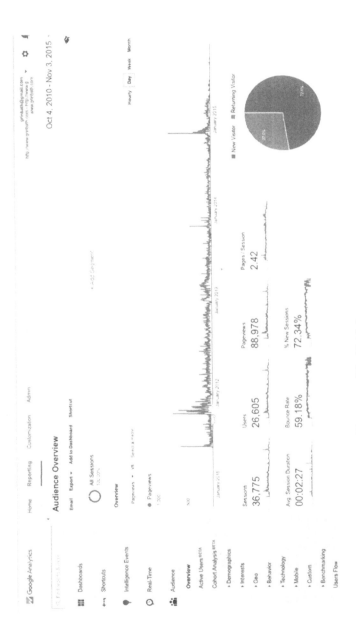

FIGURE 4.3 Google analytics for Grinbath.com. (Courtesy of EyeGuide.)

details how Americans use smartphones. This research, easily accessible from the Internet, provides us with insights about how users are engaging with these devices. If we are designing or working with designers to create an app, understanding how users "use" the smartphone devices (or tablets for that matter) will inform the various nuances of user engagement. Many other sources for statistics and reports, particularly government research entities, such as the National Institute of Health (United States) or Office for National Statistics (United Kingdom), regularly research, report, and distribute reports to the public.

Finally, do not underestimate the research your organization or your client may already possess. Sometimes clients don't even know they have valuable research about their users, so you need to inquire. Market research, for instance, can be very valuable in understanding a potential user. Market research is a quantitative analysis of a product's potential audience. Other data may include product demographic data from customer satisfaction surveys, product registration data, and even customer service logs. In Figure 4.4, we see an example from the EyeGuide website showing more user data, focusing on the activity of just one page.

4.1.2 USER FORUMS AND CUSTOMER SUPPORT

User forums are designer-managed community spaces that allow users and designers to dialogue about a specific product. Adobe and Apple, just to name a couple of organizations, use these types of forums, and their users are active participants in the space. Users also create their own forums to share ideas, troubleshoot problems with products, and even complain. In these open forums, you can find insights into common uses, problems, and even attitudes about the product.

Another tool to understand problems a design may encounter is user reviews of your product's current model or reviews of its competition. Many consumers use customer reviews and feedback frequently. Amazon, for instance, knows the importance of peer feedback. Every time you make a purchase from Amazon, Amazon requests a rating and review. Why? Because consumers read reviews before making purchases. If a product gets five stars, the consumer is more likely to feel comfortable buying that product. So, why not use these reviews as a way to get feedback about products that can inform you about users' needs, problems, or successes?

In a similar vein, you can use an organization's internal help ticket systems or customer service logs to learn about the problems users experience with existing or previous product models. These documents provide direct access to problems faced by users and solutions given to resolve the problem (we hope). This information can be invaluable when deciding how to redesign a product.

Analytics, internal survey and interview results, industry reports, data from others in your product's field, other products, assumed user communities, and even scholarly articles and other secondary resources can be highly valuable resources to include in your research. These resources not only provide the broad overview of "current" user experiences, but also ought to be used to decide on the next steps of user research, such as the kind of primary user research to conduct and, more specifically, the kinds of questions to ask.

← Metrics

Overview

Hourly   Daily   Weekly   Monthly

Introducing the medical visual tracking package.

Intoducing the medical visual tracking package.

Implement the new tandem visual surgical training system (TVST) with the medical visual tracking package from EyeGuide. Each package includes two (2) mobile tracker systems, two (2) cases, and access to our team of eye tracking implementation and training experts.

Learn about eyeguide medical training

Some places where EyeGuide technology is used.

2,796 avg 310	7,674 avg 852	2,338 avg 259
Visits	Page views	Audience size

	Visits	Page views	Audience size
November	42	108	39
October	343	908	300
September	432	1,166	371
August	335	847	274
July	292	710	243
June	318	986	270
May	250	682	222
April	507	1317	411
March	277	950	208
February	0	0	0

FIGURE 4.4 Web page visitor data for EyeGuide. (Courtesy of EyeGuide.)

4.2   OBSERVATION

Observation methods take more time, but they may be some of the most useful tools in understanding user workflow, tasks, and environment. Observation methods require that you embed yourself in the users' environments to understand the intricacies of their work and how the design (existing or developing) fulfills the needs of their tasks. Observation helps researchers understand the use scenario for designs and future testing and provides a realistic view of the conditions in which the design will be used. Whereas self-reporting, which is discussed in the next section, relies on users' perceptions of a given situation, observation allows the designer to see firsthand what happens when users interface with different designs. Observing user experiences will be a pivotal point of your research and design.

4.2.1   Site Visits

The site visit may be the most intensive of all the methods presented here, but it provides context that otherwise could not be garnered. Site visits are a type of contextual inquiry focused on "getting the best design data while involving and immersing the [design] team in the lives of their users" (Holtzblatt and Beyer 2015, 11). By observing the use environment, you can truly get a sense of what happens in that environment on a day-to-day basis.

Site visits should be performed early in the design discovery stage and at least two to three users should be observed. You want to report on three areas:

- User analysis
- Task analysis
- Environment analysis

User analysis means learning everything you can about what users do, think, and say. Further, you want to gather user demographics. What you learn from site visits about users should go a long way in defining users and their wants, needs, likes, and dislikes.

Task analysis means documenting every task and subtask performed—not instructions for doing something, but tasks that users have to think about and carry out in their environments to get work done.

Environmental analysis means taking note of everything, such as noise levels, light, size of the work area, objects in the area, resources used for support, interruptions that occur, etc. Try to bring back artifacts from the site visit (e.g., photos, maps, layouts) that illustrate your users' environments. Once you've completed your site visits, you should be able to have constructive conversations about your users.

There are several options for conducting a site visit:

- User shadowing: Follow users and take notes on their tasks, difficulties with those tasks, and environmental factors that influence task completion.
- Active intervention: Ask users questions as they perform their tasks. This technique works well when they are heavily engaged in their work and

think-aloud protocol would be too taxing. Intervention questions create open-ended explanations and may include, "What's next?" and "Why did you do that?"

- Post-task interview: Ask questions after your user completes a particular task. This gives you time to observe users uninterrupted, take notes of their actions, and have a conversation about workflow, errors, complications, or successes.
- Think-aloud protocol: Ask users to talk continuously, explaining the steps they are taking and their decision-making process throughout the task as they are completing it. This method from psychology is strong because users have limited time to contrive or make up thoughts (which can happen, unintentionally, in self-reporting methods). Further, users' short-term memory can reveal how they mentally model or think about carrying out tasks.
- Critical incident reporting: Ask users to explain everything they would do if there were a critical incident, such as a software or task error, preventing users from easily completing their tasks. If you can't observe as users work or if the task in question isn't done every day, this information can provide insight into how failures are resolved.
- Cued or retrospective recall: At the end of the site visit, discuss specific choices users made while completing tasks or dealing with problems. Remind them of their specific choices, and prompt them to describe how they chose the actions performed.

To conduct a site visit, you must first get permission from your users and their organizations. Let both know how long you will observe, what you will observe, and what methods (described above) you will use to observe. Decide first on your goals for the visit—what do you want to learn? After that, determine the method(s) you'll use to observe before you arrive. You can use more than one method, but be cautious of interfering with your users' work. Not only is this considerate to them, but it also helps you see how they work in nonintrusive situations. Have all documents needed for the site visit prepared beforehand, including consent forms, surveys, and interview questions. Finally, choose a method for recording your observation. Perhaps the simplest method to record information is using a blank piece of paper for notes or creating an observation log such as Figure 4.5.

Alternatively, you can record portions of the site visit using audio or video equipment, which should be ready to use when you arrive; this requires permission from the users and their organizations as well. If using these devices, always have a second method of data recording—batteries die as do products. A hard-copy backup will save you problems in the long run.

Task/incident	Time	Comments

FIGURE 4.5   Observation log.

At the conclusion of this chapter, a sample site visit is prepared, which provides an example of using think-aloud protocol, active intervention, and observation to understand a system and its users while completing a task. In this example, the designer, Maggie Waldron, observes a geosciences teacher at a public high school in Chicago, Illinois, to better understand how she incorporates technology into her teaching and how, for future use, the website that the designer is responsible for maintaining might create content or other features to serve the needs of this teacher. Notice the careful attention to detail in how users are profiled, the environment is described, and how user tasks with technology are analyzed. The designer carried out more than one site visit, each time with a different user, to provide comprehensive knowledge about each person's wants and needs as expressed in context. Without understanding how users used technology as part of teaching, putting up support for it on a website would just be guesswork. This is true not just for this product, but for anything you design for users. The site visit is time-consuming but a thoroughly comprehensive method for understanding users.

4.2.2   FLY ON THE WALL

Although the site visit is a terrific method for collecting comprehensive data about user experiences, your presence—talking to users and watching them work—may cause them to become self-conscious, and this self-consciousness, in turn, can cause them to work differently. The fly on the wall, however, watches but does not hover. The benefit to this is observing users whom you may not be able to otherwise (because of access or permissions).

Essentially, when you choose to be "a fly on the wall," you try to blend in to users' environments to observe user characteristics, their environments, and how they use tools in those environments. Being a fly on the wall means trying to be invisible to users; they don't know you are there and, what's more, don't care. For example, Brian's lab team used the fly on the wall technique to observe students using a scanning machine called *Bookeye*. *Bookeye* was primarily designed for scanning books, journals, diagrams, and sketches. However, while conducting the observation, researchers noticed that most users consisted of students who scanned lecture notes from their classes and then emailed those notes to themselves and, perhaps, classmates who may have missed class on the day the notes were written. Thus, *Bookeye* was being used for an entirely different purpose than it was intended. If researchers had chosen a site visit in this situation, rather than be a fly on the wall, students may have been aware of *Bookeye*'s purpose, which could have influenced their use of it.

The fly on the wall method is not only useful for understanding user experience early on in a project, but also helping during various stages of discovery and product testing by revealing needs that other products haven't met yet, observing users interacting with competitors' products and how they want to use these products, and even observing users interacting with prototypes during the design stages. Researchers do need to be careful to avoid their own confirmation bias or using observations to confirm existing beliefs. To avoid this bias, using multiple observers and/or following up observations by collecting verbal data (such as informal interviews) can be used to cross-check your observational data.

4.2.3 SHADOWING

The user shadowing method falls someplace in between site visits (which includes a high level of user engagement) and fly on the wall (a nonintrusive observation method). We use shadowing when we want to understand users' thought processes and behaviors as they perform tasks. Jen Recknagel (n.d.) suggests the use of user shadowing "when exploring a research domain to gain a rich understanding of user/customer/employee motivation and to capture what people do and not what they say they do" (par. 4). Therefore, when shadowing users, be prepared not only to follow them wherever they go, but also perform the same task(s) to experience the work.

You need to understand your representative users and have a goal for shadowing them. You may find out through secondary research or a survey, for instance, who your target users are and how they would use your design. At this point, you may even have a working hypothesis for how the product will be used and what you may want to test in the long run. With this information, you can choose the appropriate environments and users to shadow.

After you have chosen the appropriate users and task/work locations, you need to gain permission from users and organizations to conduct your research. Consider creating a research team in which more than one researcher shadows multiple users. Doing so will allow for more user and contextual data collection, allowing your team to compare data to get more accurate, actionable results. Select an appropriate length of time to shadow your users. Shadowing can span a day or weeks; this time span will largely be determined by the number of resources you and your team can spare (e.g., How much time can you dedicate to shadowing? How much does it cost to travel to the location? How much time will your employer/organization allow you to research?).

When shadowing users, you want to capture as much detail about what they are actually doing. Try to mimic their tasks if not simultaneously then immediately after they perform their tasks. You may not be able to take notes while shadowing; in this case, be prepared to take detailed after-action notes immediately following the shadowing (or immediately after performing tasks). The following questions are good note-taking prompts while shadowing your users:

- What do users have to do?
- What do they want to do?
- How do they perform their tasks?
- What knowledge about the tasks do users need before performing them?
- What aspects of the tasks are simple or difficult?
- How do users use products, or how would the products be used to perform particular tasks?
- What do users navigate, confront, or mitigate while completing tasks? Are there distractions? Is the user multitasking?
- Do users like the tasks?

It is appropriate to ask users questions, but wait until after a task or process is completed so you do not distract users while they are working. You do not want to lose insights about how users perform tasks by interrupting them midtask.

4.3 EMOTION ASSESSMENT

How users respond emotionally to products has a crucial impact on their adoption of it and, after that, their use of it. As you research your users, it is easy to devote attention to their experiences, their demographic backgrounds (e.g., education), or environmental factors that will influence how they use what you make. But we cannot forget emotion.

At a large, public university in the Southwestern United States, administrators made the decision a few years ago to transition their registration and financial management software to a different vendor's product. In some cases, staff that had been employed for decades quit or took early retirement because they couldn't accept, emotionally, the idea of trying to learn what they regarded as a more complicated technology. They left before they even learned if the new system was really harder. Emotionally, they felt so negatively about the idea of the new system that they chose not to use it.

The fact is that users make emotional decisions about products all the time. The challenge for you is to find constructive ways to include emotional assessments into the knowledge you are gathering to create profiles and personas of your users. To help you with this, we suggest three methods:

- Emotion heuristics
- Enjoyability survey
- Product reaction cards

Again, this is, like the collection of methods throughout this chapter and book, not an exhaustive list of possible methods for understanding and accounting for emotion, but it is a useful start.

4.3.1 EMOTION HEURISTICS

De Lera and Garreta-Domingo's "Ten Emotion Heuristics" (2007) is an excellent resource for evaluating user affect. Their emotion heuristics include different non-verbal actions:

- Frowning
- Brow raising
- Gazing away
- Smiling
- Compressing the lips
- Moving the mouth
- Expressing vocally
- Hand touching the face
- Drawing back on the chair
- Leaning forward on the trunk

Each of these actions corresponds with an emotional response. For instance, "leaning forward and showing a sunken chest may be a sign of depression and frustration with the task at hand" (3). In a perfect setting, one in which users have positive emotional reactions to a product, the primary emotion to be noted is likely a smile. But if there are problems, especially if users are involved in doing something like carrying out a specific task or just reacting to a new product's interface, then the users' emotions will visibly reflect negative feelings.

Brian and his lab have used De Lera and Garreta-Domingo's (2007) approach before when evaluating ebooks or online textbooks for university courses. Numbers, such as mouse clicks or time to complete tasks, among others, can identify problems. But aside from users verbalizing frustration, there are not many tools to rely on to understand, through observation, their emotional reactions. But if you are testing, for example, a new product and have an early prototype of it, you can ask users to use it to do something. While they work to complete the task you've given them, make note of their physical reactions. If they express negative reactions as they work, then your next step is to determine why. It could be that users performed poorly interacting with the product because they had a negative reaction to the size of text or it reminded them of another product they don't like. Assessing emotion using these heuristics probably won't be helpful if you aren't asking users to do or react to something, but when used alongside other data analysis, it can help you get a good sense of the affect your product has on your users.

4.3.2 Enjoyability Survey

For a long time, researchers have made use of a popular, and reliable, satisfaction survey, the System Usability Scale (SUS). You can still use it, and we talk about it in Chapter 8, for determining user satisfaction with a product. But in 2013, we began to experiment with an enjoyability survey that Jeremy Huston, Greg Gamel, and Brian created, which focuses more on measuring users' emotional reactions to a product. We felt that the SUS survey was less effective when applied to game playing, for instance. The Enjoyability Survey (Figure 4.6) is comprised of 10 prompts that focus

		Strongly disagree			Strongly agree	
		1	2	3	4	5
1	This game acted as I expected.	O	O	O	O	O
2	This game is frustrating.	O	O	O	O	O
3	This game's visuals are pleasing.	O	O	O	O	O
4	This game is too difficult.	O	O	O	O	O
5	Most people could learn to play this game quickly.	O	O	O	O	O
6	This game is unfair.	O	O	O	O	O
7	Decisions matter in this game.	O	O	O	O	O
8	This game is too complex to enjoy.	O	O	O	O	O
9	I like that this game has unexpected moments.	O	O	O	O	O
10	I would like to play this game again.	O	O	O	O	O

FIGURE 4.6 Enjoyability survey.

on factors more reflective of users' emotional states, such as "this game is frustrating" or "the game is too complex to enjoy."

You can remove the word "game" and replace it with another word, perhaps the proposed name of your new product. You also can choose to test an existing product. After all, you're interested in understanding how users feel, emotionally, so that you can ultimately design a product that doesn't run counter to those emotional feelings.

4.3.3   METHOD OF SCORING FOR THE ENJOYABILITY SURVEY

For each prompt, users choose a rating from strongly disagree (1) to strongly agree (5). Initial responses are scored as a value of one. Each question is assigned a multiplier: zero at the low end, four at the high end. Some responses carry more weight than others. Prompt 10 ("I would like to play this game again") functions as a final summative assessment, allowing users to take a holistic account of their experience. Prompt 10 is more important as a result, and it is weighted almost double the other scores. As with the other scores, the multiplier for the most negative response is zero. Any other response is scored thusly: $(2 \times x) - 1x$ equals the score indicated on the Likert scale response.

After each response is scored and multiplied, the points are added together to produce a preliminary enjoyability score. This number ranges from 0 to 40, which can be a hard range to digest. By multiplying the preliminary enjoyability score by 2.5, however, the score ranges from 0 to 100, a more approachable and understandable expression of the data. Obviously, a higher score, closer to 100, means users enjoy the product more. A lower score means they do not, which would prompt you to ask users questions about why and help you understand the causes behind their negative, emotional reactions or lack of enjoyability.

4.3.4   MICROSOFT PRODUCT REACTION CARDS

Another approach for assessing user emotion is the use of Microsoft product reaction cards (Benedek and Miner 2002). Figure 4.7 shows all 118 cards.

Benedek and Miner created product reaction cards because they felt user desirability (did they want or desire to use a product?) was not being measured by other methods. Sixty percent of the cards have positive words, and 40% have negative words. After users complete a task, such as answering a question you pose to them as part of a scenario, using your new product or an existing product to do something, you place all 118 cards (buy 3 × 5 notecards or make your own, one word per card) in front of the users and then ask them to choose all of those cards that describe their experience. You then ask users to narrow down the card selection to a top five. From there, you talk to the users about their reasons for choosing the selected cards.

4.4   SELF-REPORTING METHODS

Self-reporting data is when users answer questions based on their experience without a researcher verifying the correctness of that information. This does not mean that self-reporting is bad data; rather, self-reporting can provide information about user motives, wants, needs, and desires. But, just as you would not use statistical data as

The complete set of 118 product reaction cards				
Accessible	Creative	Fast	Meaningful	Slow
Advanced	Customizable	Flexible	Motivating	Sophisticated
Annoying	Cutting edge	Fragile	Not secure	Stable
Appealing	Dated	Fresh	Not valuable	Sterile
Approachable	Desirable	Friendly	Novel	Stimulating
Attractive	Difficult	Frustrating	Old	Straightforward
Boring	Disconnected	Fun	Optimistic	Stressful
Business-like	Disruptive	Gets in the way	Ordinary	Time-consuming
Busy	Distracting	Hard to use	Organized	Time-saving
Calm	Dull	Helpful	Overbearing	Too technical
Clean	Easy to use	High quality	Overwhelming	Trustworthy
Clear	Effective	Impersonal	Patronizing	Unapproachable
Collaborative	Efficient	Impressive	Personal	Unattractive
Comfortable	Effortless	Incomprehensible	Poor quality	Uncontrollable
Compatible	Empowering	Inconsistent	Powerful	Unconventional
Compelling	Energetic	Ineffective	Predictable	Understandable
Complex	Engaging	Innovative	Professional	Undesirable
Comprehensive	Entertaining	Inspiring	Relevant	Unpredictable
Confident	Enthusiastic	Integrated	Reliable	Unrefined
Confusing	Essential	Intimidating	Responsive	Usable
Connected	Exceptional	Intuitive	Rigid	Useful
Consistent	Exciting	Inviting	Satisfying	Valuable
Controllable	Expected	Irrelevant	Secure	
Convenient	Familiar	Low maintenance	Simplistic	
(Developed by and ©2002 Microsoft Corporation. All rights reserved.)				

FIGURE 4.7 Microsoft product reaction cards.

your entire research methodology, you should also balance self-reporting data with other research methods.

4.4.1 Focus Groups

A focus group is a group of users invited by a company, designer, or research team to provide feedback about a specific topic. Essentially, a focus group should unveil user opinions, attitudes, beliefs, and assumptions about the topic in question. The purpose of a focus group is not to have all users agree on the perfect product design; rather, it will provide several viewpoints and opinions for researchers to consider before beginning initial designs. Focus groups ought to include diverse, representative users to get the best feedback. No more than 12 people should be included in a focus group session although there can be several individual sessions covering the topic (Usability Net 2006).

A moderator asks the focus group open-ended questions to get ideas about their use of a product, what they want to get out of the product (i.e., what tasks must they be able to perform with the product), their needs and wants for the product, and may even show videos about or prototypes of the new product to the focus group to illicit feedback. Because focus groups are small in scope, they are not meant to represent the entire population. However, they can help designers "hypothesi[ze] for further evaluation and user validation using both qualitative and quantitative methods" (Usability Net 2006). The opinions expressed during these group sessions can also be incorporated in initial designs that are tested using different methods.

4.4.2 SURVEYS

Surveys are sets of questions (closed- or open-ended) that give you a broad sense of user demographics and preferences. Surveys are useful in that you can get user feedback in a relatively short amount of time. However, problems can arise, such as a small return rate on surveys, survey participants that are not representative users, and surveys that are written poorly and do not accurately measure the users' experience.

Distributing and collecting surveys is no easy task, but it can be made easier by knowing how many surveys you need returned to represent your sample population. A return rate is the number of surveys sent out divided by the number of surveys returned. A simple calculation of how many surveys you need returned is determined by three factors: sample size, margin of error, and confidence level. The sample size is the number of people you will distribute the survey to. Margin of error refers to how many responses may be different from the total population. For instance, Hughes and Hayhoe (2008) explain, if my margin of error is ±5, "the researcher could say, 'I think the real value in the population falls within the reported value plus or minus this margin of error'" (103). You can calculate the sample size on your own in Excel; however, there are many survey calculator tools online or embedded in online survey software to make the process easier.

Locating your users' networks can help you reach representative users. You can have the highest return rate possible of your survey, but if you are not getting data from representative users, you are pretty much wasting your time. However, if you find professional associations, discussion boards, or wherever the users are, you can target those networks for your survey sample. Luckily, you have several options for distributing surveys. If you want to send surveys to a large organization, perhaps through a listserv or newsletter, you can embed a link to an online survey. Survey formats offered by SurveyMonkey, Qualtrics, Google Forms, and Question Pro (just to name a few) are either free or affordable options that accommodate most researchers' needs. In cases in which your users may not be as computer literate, you can still print and distribute surveys at the users' location and ask for them to be returned by mail or to a collection box where the surveys remain anonymous. Still another option is to mail out surveys and ask participants to return surveys by mail. The print option is more expensive, and there is no guarantee that surveys will be returned. However, if you know the participants are more likely to respond in print than by an online survey, this may be the best option.

A survey needs to be written well, and the questions asked need to accurately measure your users' demographics and preferences. Surveys consist of two types of questions: closed- and open-ended. Closed-ended questions are basic yes/no questions or those that provide answers (i.e., a multiple choice format). An example of a closed-ended question is a survey that asks, "How likely are you to use an app that can detect concussions?" and provides options to select on a Likert scale (on which 5 is very likely and 1 is not very likely); you are asking participants to choose one of the given options. Closed-ended questions have the benefit of being quantifiable and easily calculated using survey software or Excel.

On the other hand, open-ended questions allow participants to answer questions in their own words. For example, perhaps you don't know all the uses for concussion detection. A follow-up question to "How likely would you be to use an app for concussion detection?" may be, "In what cases would you use this app?" Instead of giving participants options to choose from, you'd provide a blank space in which they can write their response. The benefit of this type of question is that users can provide information from their own perspectives and give you ideas you've never thought of. The disadvantage is that these questions are more difficult to code because these questions provide qualitative research. Instead of being automatically generated in SurveyMonkey (see Chapter 15), you would need to read all answers and code or categorize participant responses. Because this takes more time, you'd want to use open-ended questions sparingly when they are really needed to understand your users' priorities (see Figure 4.8 for survey question examples).

Finally, pilot test your survey before distributing it to participants. Find a couple of representative users who can pilot test the survey. Not only will this help you determine whether the survey is reaching the desired users, it can also reveal unclear,

Closed-ended questions	Open-ended questions
1. How likely are you to use a mobile app to deposit a check? (1 = not at all likely, 5 = extremely likely)	1. What's your opinion of mobile banking apps? (ex: "I don't think they're safe.")
2. How many times per week do you use text messaging apps? (A = 0, D = more than 50)	2. Under what circumstances do you normally send a text message? (ex: "When I don't have time to call.")
3. What is your attitude towards learning a new technology? (1 = hate it, 5 = love it)	3. How would you describe your attitude towards learning a new technology? (ex: "It depends. If it's something I really want to learn, I'm okay with spending more time on it.")
4. When do you typically shop online? (A = 7 am–9 am, D = after 10 pm)	4. When do you typically shop online? (ex: "Whenever I can! Normally during my lunch break, or after the kids are asleep. I need a good 20- or 30-minute chunk of time to select the items and check out.")

FIGURE 4.8   Example survey questions.

biased, irrelevant, or unanswerable questions. These tests can also make you aware of questions you need to add to your survey.

4.4.3 INTERVIEWS

Interviewing users, formally and informally, is a quick and easy way to understand how they perceive their use environment, their frustrations with the usability of designs, their motives or purposes for using a particular product or design, and their own perceived needs. Interviews are convenient in that, when you have access to representative users, they can be done quickly and even spontaneously. When researchers use interviewing as a method, they tend to conduct fewer interviews that are more in depth about the user experience than, say, surveys or focus groups.

Interviews can be focused on finding various sets of data at any phase of design development. For example, user interviews can be used to share undeveloped concepts to see how users feel about some of your initial project ideas and if the users would interact with any of the designs. Users can also be interviewed during the rapid prototyping stage. You can run an informal usability test with individual interviewees and ask them what they think about the prototype using questions, such as "Would you want to use this? Why? How often would you use it? In what circumstances?" With informal interviews, you can use feedback that you received from previous users and create additional interview questions that will help you get comparable opinions regarding your designs.

Be aware that what users say may not correlate with what they actually need or desire in a design. Don't get caught up in trying to appease all users' individual preferences. Further, if users aren't fully representative, you might lead your design astray by taking user feedback too literally. You can avoid this by screening your interviewees to make sure they are representative users and, before the interview, give participants a clear idea of the use scenario (i.e., the context of use).

Unlike surveys, interviews are almost always qualitative. Instead of looking for cut and dried solutions to design preferences, you should pay attention to patterns that appear across your interviewees. Further, before making concept design changes or applying changes to prototypes, confirm what the interview data implies with other secondary and observational data.

4.4.4 DIARY STUDIES

It is not always possible to observe users for a long period of time. However, having insights and records into users' activity over an extended period can capture user processes and problems not seen during a short-term observation. Diary studies can be useful in this case. Kuniavsky (2003) defines diary studies as "a group of people keep[ing] a diary as they use a product. They track which mistakes they make, what they learn, and how often they use the product" (369). Diaries should be collected from more than one participant so that results can be cross-referenced and patterns between participants can be observed.

Diaries can be structured or unstructured. Some of these forms may even be used together. A structured diary is a document created by the design team to be completed by the user. For instance, users may be asked to fill out a form every day that asks about how they used the product, how often the product was used, what tasks were completed using the product, and what problems occurred while using the product. This structured diary is geared toward providing designers with need-to-know information. An unstructured diary, as you can guess, asks users to write about their experiences without the additional prompting a form would provide. Even in this case, however, users need to be given a clear focus for the type of content they should include. Unstructured diaries may lead to tangents in the user's entries; however, these tangents may provide additional information about the product that structured diaries would not allow for (Kuniavsky 2003).

Diaries can also be kept using various media, such as paper, audiovisual, and even mobile apps. The most traditional media is the paper diary. It can be the journal we are most familiar with or a packet of forms you provide the user. However, other media can also be used. Users can diary their experiences using an audio device, which can capture both the day's process and users' tones and attitudes as they describe or recall their experiences. Video can also provide the same benefits of an audio recorder while also capturing pictures of equipment, errors, or successes. A newer form of diary keeping is mobile apps. Mobile diaries are meant to free users from a static work environment, understanding that much work is done on the go. Thus, users can easily track events of the day by using their smartphones. In addition to convenience, mobile apps, such as Punchcut's new mobile app, uses inherent features of a smartphone to collect data such as using audio recording functions for voice capture and the phone's camera to collect images. This data can then be sent directly to the researcher (Sparandara, Chu, and Enge 2015).

A mixed-tools approach can certainly be used to gather diary data. Some researchers create design probe kits that may include "diaries, question cards, postcards, disposable cameras, or other tools for mapping or drawing" (Helen Hamlyn Centre for Design n.d.). The purpose of a design probe kit is the same as any diary study: get information about how users perform tasks, work, or live without the interference of researchers. For example, Kate once participated in a study that aimed to see which contextual and environmental factors influenced the writing process. The researcher wanted to capture video of actual work as well as reactions to that work. Kate installed screen, video, and audio capture software on her computer, and every time she sat down to write, the software was turned on to capture activity. In addition to capturing how she was writing, she also could discuss her frustrations while using word processing and research tools as well as her attitude about the writing project from the audio and visual elements included. The researcher prompted Kate via email several times to continue capturing, saving, and sending the data so she could analyze the process. After the study's time span was completed, the researcher analyzed the materials and interviewed Kate to collect additional data about her own workflow perceptions and attitudes about writing tools. This method allowed Kate, the participant, to work in her own environment with her own equipment without a designer or researcher hovering. Yes, she still knew that the material would be reviewed, and this could have impacted her performance (a danger of all self-reporting methods); however, the tools used to gather her data were less intrusive than having an observer watching Kate while she worked.

Diary entry	Interview question
1. Used skype at 9 am to call Cindy.	1. How long did you talk? Where were you? Did you have any connection issues? Did you use video or just audio?
2. Deposited check from Grandpa at 2 pm while waiting at the dentist. Took about 3 minutes.	2. What app did you use? Did you successfully deposit the check? What surface did you place the check on?
3. Downloaded new songs on Apple iTunes at 4 pm. Took about 6 minutes.	3. Where were you? How many songs did you download?
4. Used FaceTime at 8 pm to video chat with Daniel. Sat on couch in living room.	4. How long did you talk? Did you have any connection issues?

FIGURE 4.9   Example diary entries and corresponding interview questions for UCD study on mobile app use.

The method or media you choose to use should be appropriate for the scope of your project, and tools should be available to users and limit intrusiveness in their work. After diaries have been collected and analyzed, you can follow-up with users in interviews (Figure 4.9) to find out more about their experiences.

4.5   DESIGNER ANALYSIS

Designer analysis methods break down processes without the input of users, perhaps because user data has already been collected and is now being analyzed in a different way. Therefore, the major difference between designer analyses and other methods is that the designer or researcher uses previously collected data or knowledge to determine user needs or processes. Here is an example: You perform a site visit at a software company, specifically observing a worker who writes documentation for software. As part of your site visit, you note all tasks as part of your data collection. After the site visit, you might decide to perform a task analysis using this data. This task analysis then becomes new information about the order of tasks from start to finish. A designer analysis can also lead to creating user journey maps or storyboarding. These methods include affinity diagrams, use cases, task analyses, and user matrices.

4.5.1   AFFINITY DIAGRAMS

Affinity diagramming is a useful tool to rely on for answering the question "What do we know?" Later on in the book, we suggest, in seeming opposition to what we suggest here, that you also can involve users in affinity diagramming at later stages of product development. The fact is that you can involve users in any activity, including actual design, but when you use affinity diagramming to research users early on in the design process, you try to understand who your users are, and so you can't really involve them yet.

When using affinity diagramming, you are interested in learning what relationships exist between ideas and concepts discovered during data gathering. You can

FIGURE 4.10   Affinity diagramming exercise. ("An affinity diagramming session," own work: Marge6914, May 8, 2014. Creative Commons Attribution 3.0, https://creativecommons .org/licenses/by-sa/3.0/deed.en.)

start with big concepts, such as user characteristics, and then further narrow your scope after successive rounds of testing. You can focus on answering broad questions (e.g., "What do users need to use this product effectively?") and then go through your data looking for affinity or relationships that make sense.

Perform this activity with your design team—a small or large group ideally comprised of those most involved in the design. Affinity diagramming is useful for a large team with multiple interests and skill sets; a team with varied purposes and interests for being part of the activity can discourage groupthink (everyone agreeing with what the rest of the group says without critically examining the situation). Limit your session to two hours or end when your group is getting tired or bored. Here is a good way to run your affinity diagramming session:

1. Find a large, flat surface where you can work; this might be a blank wall, whiteboard, large window, or table (Figure 4.10).
2. Search through data individually. Make note of each piece of relevant data (words, images, numbers, etc.).

3. Write down one thought, idea, issue, or observation per sticky note.
4. Have the group begin sorting sticky notes into groups or categories. Make sure all participants are involved and one or two people are not dominating the activity.
5. Limit each group to containing five notes. After five notes, start a new group.
6. Actively discuss where notes should go. Not everyone will agree about note placement—that's okay. The point of this activity is to discuss how data relates to the team's different perspectives. Discuss as a group where notes go and how they relate to other items in the group. If the group does not come to a consensus, duplicate the note and put it in both groups.
7. Create new notes if new ideas and observations come up during sorting and discussion.
8. When all notes have been grouped, come back together to label (or characterize by theme) each group (Information and Design 2012).

What is ideal about affinity diagramming is that it encourages the synthesis of ideas and concepts without privileging any one participant in the process. You work with your team to create affinity groups or chains based upon what you see. No one can tell you what you see. You look at the data, identify affinity, and other members of your team do the same. If affinity exists, categories will form. If relationships are not revealed, then the groups identified will not be included.

You'll have a lot of data, and that may make the task of synthesizing it into design priorities (e.g., what information matters most, what our user profiles are, product requirements lists, overall design) difficult. Affinity diagramming helps, in a more organized and arguably constructive fashion, to collect together and analyze a lot of diverse ideas about your users.

4.5.2 Use Cases

Use cases document how users perform within their environments using the tools afforded to them. There are two main parts to a use case. First, a use case notes users' goals and describes tasks until the goal is met without specifying the user interface. Second, a use case determines how the design should respond to users' actions (Usability.gov n.d.).

To write a use case, you must first identify the actors in the system (such as who is using the interface/product), the interaction or what users want to do, and the goal for using the system. Often, use cases are written as easy-to-understand narratives but can also be composed as outlines, tables, and even visual models. Kenworthy (1997) provides an eight-step process for writing a use case:

1. Identify who is going to be using the interface.
2. Pick one of those actors.
3. Define what that actor wants to do using the interface. Each thing the actor does with the interface becomes a use case.

4. For each use case, decide on the normal course of events when that actor is using the interface.
5. Describe the basic course in the description for the use case. Describe it in terms of what the actor does and what the system does in response that the actor should be aware of.
6. When the basic course is described, consider alternate courses of events and add those to "extend" the use case.
7. Look for commonalities among the use cases. Extract these and note them as common course use cases.
8. Repeat steps 2 through 7 for all other actors (Kenworthy 1997).

The use case can be simple. For example,

> The instructor spends three hours every evening grading assignments. To grade assignments, the instructor accesses them in Blackboard. The instructor grades an assignment. The grade is recorded in the grade field box. The instructor moves on to the next assignment.

You could also create a more specific use case that breaks down each step the actor (instructor) goes through to reach the goal (grade assignment). Because the latter is more structured, you might use a table to indicate each task and, perhaps, interferences when completing the task. Figure 4.11 demonstrates how a more detailed use case may be reported.

Use case	The instructor grades student assignments using blackboard.
Actor	Instructor
Interface	Blackboard grading module
Description	The instructor spends three hours every evening grading assignments. To grade assignments, the instructor accesses them in blackboard. The instructor grades an assignment. The grade is recorded in the grade field box. The instructor then moves on to the next assignment.
1	Log in to blackboard and navigate to grade center. Choose "needs grading option."
2	First student assignment is selected and opened in "view attempt" screen.
3	Instructor uses blackboard feedback tools to comment on the assignment.
4	Instructor reviews feedback comments and uses rubric (a separate file from blackboard) to determine assignment grade.
5	Instructor writes a summary of feedback comments that use rubric language in blackboard comment box. Instructor fills out rubric and copies and pastes it in comment box.
6	Instructor puts assignment score in grade field.
7	Instructor submits assignment score.
8	Instructor finished grading one student assignment.

FIGURE 4.11 Use case example.

Figure 4.11 could include even more detail and be broken up into several processes. For instance, a separate use case could look at using just the commenting tools to give students feedback. These might describe when the instructor uses comment boxes versus strikeouts versus highlights.

4.5.3   TASK ANALYSES

A task is "any observable, measurable action that has an observable beginning and an observable end" (Miller 1997, quoted in Hackos and Redish 1998). Even the simplest of processes include several steps. Kate recently learned how to use Snapchat, a social media app with which you snap a picture and send it to a friend or group of friends. Sending her friends a photo through Snapchat requires several steps. First, she opens the Snapchat app on her smartphone. Next, she snaps a picture. Then, she adds a caption to her snap. Next, she clicks the arrow that takes her to her contacts screen. From the contacts screen, she selects recipients of her snap. Finally, she presses the arrow on the bottom right-hand corner—the end of the task. Her friends will now receive her snapchat photo. This is one task from start to finish.

Just knowing the series of steps that make up a task does not tell us much. This is why we need task analyses. The purpose of a task analysis includes three aspects:

1. Learn about your users' goals and their ways of working.
2. Break down tasks into smaller steps to show what users must do to meet their goals; understand users' knowledge and experiences that inform how they complete tasks.
3. Understand users' physical, social, cultural, and technological environments.

There are two types of task analyses: procedural and hierarchical. According to Hackos and Redish (1998), a procedural analysis "take[s] one specific task and divide[s] it into the steps and decisions that a user goes through in doing that task" (75). You begin with the first step (the start) and end when the task has successfully been completed. Figure 4.12 illustrates a procedural task analysis of a user's observed tasks while using a content management system (CMS) grading component.

You'll notice that simple tasks are represented in boxes, tasks that require decision making are in triangles, and start and finish points are in circles (you will notice a similar pattern for user flows). The point of the procedural analysis is to see what exact steps are or will be taken and in what order. Not all flowcharts will look as linear as Figure 4.12. Some may include arrows that link "main screen" to "assign grade" in the event that the user has left and come back to grading after a period of time. Or the user may think of more comments to add to the assignment while filling out the rubric, in which case an arrow could be drawn from "fills out critical thinking rubric" back to the comment triangle.

The second type of task analysis is a hierarchical analysis. Seels and Glasgow (1990) explain, "A hierarchy is an organization of elements that, according to prerequisite relationships, describes the path of experiences a learner must take to achieve any single behavior that appears higher in the hierarchy" (94). In this case, the higher the action in the hierarchy, the more essential the action is to completing all

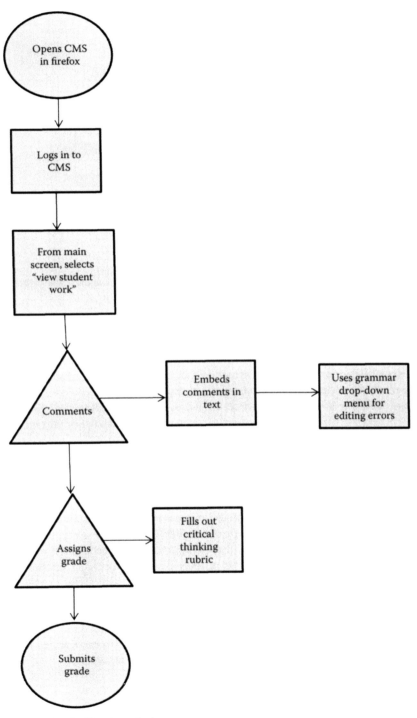

FIGURE 4.12 CMS task analysis.

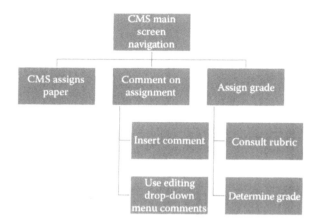

FIGURE 4.13   Hierarchy task analysis.

lower-level tasks in the hierarchy. If we were to use our previous user's task sequence again, our hierarchy may resemble Figure 4.13.

4.5.4   User Matrices

A matrix is a way to structure data in such a way that it shows patterns or can be used to develop new data. You may recall matrices from math or business courses. For user research, we can use a user matrix to categorize data based on user characteristics observed or found through other methods. Hackos and Redish (1998) note that the characteristics you choose to include in your user matrix should be related to the goals of your design.

How you design your user matrix will be determined by the data you have collected. If you are creating a user matrix from site visits, you may brainstorm with your design team the different characteristics of the users observed. If you have survey data, you can use demographic and behavior data to create your matrix. In either case, you need to first determine the important user information needed to build your user profiles. Figure 4.14 provides several examples that can be used to create your matrix; however, not all of these characteristics need to be used. Only those necessary to understand the context of the project are needed. Therefore, prioritize the characteristics before brainstorming, analyzing, and creating content for your user matrix.

As an example, Kate did a study on how students use course syllabi. Before conducting testing, she sent out a survey to determine her user population's characteristics (see survey at the end of this chapter). Based on this survey data, Kate narrowed the main characteristics of syllabi users into the following categories:

- Gender
- Age
- Area of study by college unit

- Nontransfer or transfer
- How students access syllabus
- Frequency of use

Kate then created the following user matrix to better understand who her user population was and how they may use a syllabus (Figure 4.15). Notice that the syllabus user matrix only identifies four user profiles. These user groups could have been extended further, but that would have created more user profiles than would be usable for beginning and testing designs. Therefore, four user groups were identified, using characteristics that were representative among different groups.

In addition to creating profiles that can help you and your design team create design prototypes, matrices can become sampling tools when you begin testing your design.

Personal characteristics	Task-related characteristics	Geographic and social characteristics
Age	Goals and motivations	Location
Gender	Tasks	User's culture
Socio economic status and lifestyle	Usage (frequency or type of use)	Workplace culture and tools
Job type/organizational role	Training and experience	Environmental conditions
Physical abilities and constraints		Social connections
Emotions and attitudes toward product (i.e., technology)		

FIGURE 4.14   Characteristics for user matrix.

	User 1	User 2	User 3	User 4
Gender	Female	Female	Male	Male
Age	18–23	18–23	18–23	18–23
Area of study	Arts and sciences	Human sciences	Agricultural sciences	Engineering
Nontransfer or transfer	Both	Nontransfer	Transfer	Nontransfer
How students access syllabus	Course management system	Course management system	Course management system	Course management system
Frequency of use	Rarely (1–3 times)	Occasionally (4–9 times)	Occasionally (4–9 times) and rarely (1–3 times)	Occasionally (4–9 times) and rarely (1–3 times)

FIGURE 4.15   Syllabus user matrix.

4.6　USER AND USE DIAGRAMS

So far, we have been encouraging you to perform user research by going into the field where work is being done, selecting representative users to provide information about use and demographics, and conducting analyses based on this information. Unfortunately, these methods are not enough to begin designing your product. Going into the field without a clear purpose yet expecting to see something happen that you know is significant while it happens may be unrealistic. There is too much to see, and often we report back with too much data to successfully analyze. This can lead to picking off only the most obvious observations and incorporating them into task development or even product design. To better focus our observations in the field, we need to know why we are going out there and what we are trying to find. Therefore, we need to emphasize discovery, even before field observations, of critical aspects related to users by asking the following questions:

- Who are our users, and what experience do they currently have?
- What competitor or previous products have been used to satisfy user needs?
- What are our resource limitations as we design our new product?

If you want to go into the field and watch users use your new product, use that feedback to improve the design, and eventually transition into full-blown user testing, you need different approaches to gather knowledge about users that will tell you more clearly what goals they have and what features are required to allow them to achieve those goals. A few methods can prepare designers for their expectations of users in the field, such as user flows, user journey maps, and storyboarding. These methods not only show you your own assumptions about users, but also set your expectations for how the interface will respond to user needs.

4.6.1　User Flows

A user flow chart can help you learn more about your users before you set up observations or create tasks for user testing. Basic information about your users might not be enough to understand what they want to do with the product. For instance, you can describe a user with a few defining characteristics, such as the following: Bob is 14–18, attends a public high school, has used Acme Mobile App Maker's products before, has an affinity for electronic devices, and has enough experience with information on mobile devices to know his way around mobile interfaces. This is great information, but it does not tell you how Bob goes about using information on the mobile interface. You might have enough information to recruit people like Bob for user research or testing, but you do not have enough knowledge to look for particular conditions as you visit Bob onsite or as you construct tasks for him when you test the interface.

A user flow chart (Figure 4.16) can help with this. Pick a process the user will have to perform, such as editing an application profile. First, have Bob start where you would expect him to start, which already helps you think about how the user will flow through the task. If your application is intuitive, the interface will allow Bob to flow through the task as he expects to.

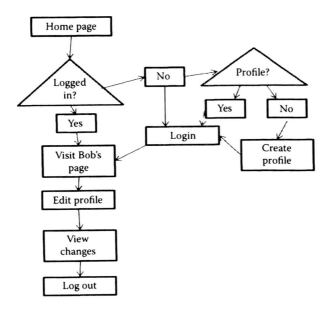

FIGURE 4.16   User flow chart.

From Bob's starting point, create squares for every page Bob would encounter along the way to achieve his goal. Whenever Bob has to make a decision, surround that step with a triangle. If multiple pages are available for Bob to choose from, put those into the flow and, if necessary, indicate where supporting files need to be accessed (Figure 4.16).

This chart can get complicated. The more complicated the flow, the more likely you are designing a difficult interface. If you find the user flow complicated, continue constructing new user flows, cutting down on the number of complications the user will encounter. Once you have narrowed down the user flow, you have a better sense of what Bob is doing. When you go into the field with this information, you will look for the bottlenecks, problem spots, or navigational pathways that you noted when you created your user flows.

4.6.2   User Journey Maps

A user or customer journey map is a visual depiction of what users need and what steps they take to fulfill those needs as they interact with a product. Simply, you create a visual that allows you to display, together, what customers are trying to do, what steps they want to take to do it, and what emotions or thoughts they experience as they carry out these steps (Flom 2011, par. 4). Creating a journey map has several benefits:

- You can continue developing your map over time as you gather more data about users.
- It can prepare you for site visits (when you will be collecting more user data).

- A map can show how the design and testing of your product resulted or did not result in creating an experience that was successful in allowing people to take the journey they wanted. This information and the visualization of this information are important because it can be easily communicated to stakeholders. You can remind stakeholders of what users want to do, why they want to do it, and what they think and feel while going through the process as well as what users cannot do.

According to Flom (2011), there are multiple factors that make a good journey map, but we want you to focus, in particular, on three: real research, real behavior, and real experiences. Note here that, unlike some other methods, "real" is emphasized. Real research, whether this be an observation or a documented user's experience from a community forum, should be detailed enough to go back to the source of information to produce your journey map and also an experience that the map's users can find credible. The user map should represent real user behavior or actions, not those we assume. Why is this important? We do user research because if we instinctively knew how users would perform tasks, all of our designs would be perfect from the start, but we know this is not the case. Again, although we can continue to modify the user map, its contents from the outset should represent observable user behavior. Finally, we want to capture the real experiences of users. This means not only showing how they perform tasks, but their emotional reactions to doing so.

Figure 4.17, "Understanding the Texas Tech University Student Bill," illustrates a wonderful interpretation of a user's journey. In this example, Kim is a new distance graduate student at Texas Tech. However, the journey map shows complications with understanding her bill and communicating with the university over tuition problems. She first notices the problem on her own and contacts the business office to find more information. The business office tells her that her financial aid was in process, but the emailed invoice from the university says she owes money. Further, the invoice notes that she will be cancelled from classes if tuition is not paid. The journey map shows us that Kim is confused by the varying communication. Further, she is worried about what will happen if the bill is not paid. As we go through the journey map, we notice more conflicting confirmation and billing notices that further complicate the problem and distress Kim.

There is much to learn from Kim's experience. First, if our stakeholder is the university's billing office, we can note the number of times the user was given conflicting information that prevented her from completing her task (i.e., paying her tuition bill). This was caused by different communication from multiple offices, undefined billing codes, and conflicting customer support resulting from communicating with several different agents in the office. This journey map does not show an ideal user experience, but it does show a realistic one. As designers, we can go back to stakeholders and recommend a new design or process that could make communication between different offices more transparent, redesign the way automated billing invoices are sent out, design a billing code guide for students, and streamline the customer service experience.

Figure 4.17 shows us problems with the existing design and the possibilities for redesigning the system to be more user-centered. Flom (2011) claims that journey

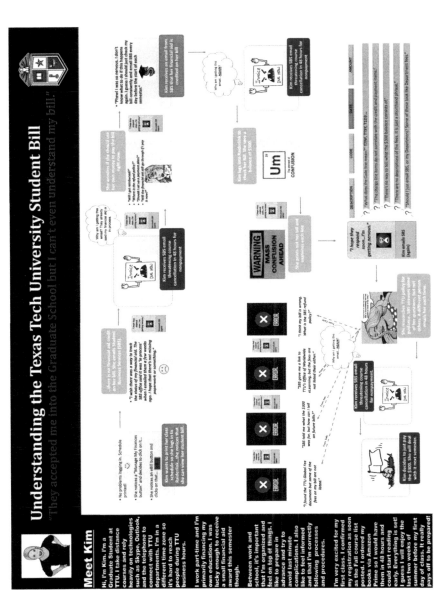

FIGURE 4.17 Understanding the Texas Tech University student bill user journey map. (Courtesy of Sarah Martin.)

maps can also be designed showing the "optimal experience" so that the real experience and the optimal experience can be contrasted. What customer journey does the designer want the user to experience? Based on Kim's journey, how do we make changes to the design so that Kim's experience is more like the optimal experience, rather than her current experience? Journey maps can help us visualize how to get there.

4.6.3 STORYBOARDING

You may relate storyboarding to the fields of advertising, television, and film in which storyboards are created for commercials and even full-length movies. The storyboard is the trusted first-stage prototype in all of these fields. It is a low-stakes way of conceptualizing and pitching ideas before investing in film, animation, or document production. Further, it serves as an outline for what the story's narrative should look like. Writers, directors, advertising executives, and many other stakeholders create storyboards or provide constructive feedback as ideas are pitched to them.

One of our arguments in this chapter has been that we must understand users' experiences, their needs, and their uses of design before investing in more expensive prototyping and design methods. The storyboard is a highly useful technique for understanding your users' narratives—their stories about what they need, how they would engage with a design, what actions they would perform, what problems they would encounter, and how their use of the product ends (which can be a success or failure). Like other methods discussed, the storyboard ought to be created using observational and self-reporting data (the data you have already collected about actual users). You may even find it necessary to create more than one storyboard: one for each representative user identified.

You can use many tools to create a storyboard: Microsoft PowerPoint (using one slide per frame), Adobe InDesign or Illustrator, and even online storyboard templates at your disposal. However, we suggest starting with a big piece of paper (you can also tape together smaller pieces, e.g., printer paper, to make it work as something larger) for drawing out an initial storyboard. As you begin drawing, picture a comic strip. Each scene, or frame, explains a key part of the user's story (words and actions) conveyed through drawings.

The user's story should have a beginning, a middle, and an end. How would users begin interfacing with the design? Draw it out. How do they feel (what emotions would be expressed) as they engage with the design for the first time? Draw it out. Each scene should represent one action or response to an action (Figure 4.18). Use the storyboard not only to tell others about a user's experience, but also to discover a user's experience. In other words, you often know how the story is going to end or you have a plan for that ending, and you use the storyboard to tell others about it so they can help you with implementing the story.

What if you don't know the ending? Use the storyboard as a constructive way to talk about what users may do given certain scenarios. If you have an idea of the users or situations they'll be in, use storyboarding to put together the things that have to be done for users to achieve key goals in those situations.

FIGURE 4.18 Storyboard. (Courtesy of Dr. Briar Still.)

Ultimately, when the storyboard is complete, the storyboard's audience should be able to follow the story and understand what the story is about. Figure 4.18 is an example of a low-fidelity storyboard.

The user starts off being told she needs to complete an online form for the Army. Not sure what to do, she begins by looking for the form on the computer. After finding the form, she fills it out online, but is unable to sign it. Again, the user questions what to do. She finds the print button and selects it. Then, she is able to sign the form and turn it in to the appropriate office. The final scene shows a successful task completed. In addition to telling us the user's story, the storyboard should show us what is missing and what else needs to be there for users to achieve their goals. For instance, in the case of the Army's print form, we may notice that the user does not know where to start or how to print the form. This is an area the design team can work on.

You can build your storyboard in six steps:

1. Collect your storyboard materials and all existing user research.
2. Choose one representative user for your storyboard.
3. Start from the beginning. What is the first interaction the user will have with the product? Illustrate this first interaction in the first frame.
4. Continue creating scenes for each major step and user emotion. Make sure you have a clear beginning, middle, and end to your storyboard.
5. If necessary, indicate the narrative's reading orientation by numbering frames or adding arrows to show the scene's order.
6. Revise the storyboard as you begin collecting more information through brainstorming, research, and design prototypes (which are discussed in Chapter 7).

Finishing a storyboard does not mean it remains static; it will change, get richer, and have more accurate scenes as you continue to study and understand the user experience. You can begin with a very rough storyboard, maybe just text. But over time, you can and should make it more realistic, adding in rough drawings of the scene's environment, the user's actions, and the user's verbalizations. When you get to the point at which you feel the storyboard is as representative of the user experience as possible, consider refining it for stakeholders to begin selling your design or redesign ideas. From here, you are ready to start thinking about ways to develop the story into a higher fidelity prototype of the product.

4.6.4 CARD SORTING

Card sorting is used to see how users would group or categorize information. This information is useful when determining the information architecture of your product, such as the organization of a website. Card sorting can tell the designer where users would expect to find information by having them sort ideas, concepts, or statements into groups.

To begin card sorting, you should brainstorm what contents you know need to be in the design and even desired components from your users (these factors can be determined from other user research). Further, you need to understand who your representative users are so that your participants will give you data that is applicable to your larger group of experts. They don't have to be from the same user group; for instance, you may choose to split up the users into a group of novice users and a group of subject matter experts (SME). Each user should sort the cards on his or her own. Determine whether you are going to use an open or closed sorting technique: open sorting has users group cards on their own with no preexisting taxonomy (or structure) and then indicate what each group of cards has in common (such as characteristics of information, etc.); on the other hand, closed sorting has users sort cards into groups predetermined by the designers. These groups may be menu headings that the designers already have created or that already exist on the website. In each case, the user should be allowed to have a "nonsort" pile or a pile in which they feel cards have no home in any created category. Once participants are finished with the activity, designers should document which cards belong in which category. To summarize, follow these steps to conduct your card-sorting activity:

1. Write each idea, concept, or statement on an index card.
2. Determine whether you want to use open or closed sorting.
3. Find at least six representative users to participate in your card-sorting activity.
4. Involve only one user in each card-sorting session.
5. Shuffle the cards so they are randomly selected.
6. Have the participant group cards into categories (either participant created or predetermined by the designer).
7. If the participant has created his or her own categories, have the participant describe what each category represents (perhaps provide sticky notes so the participant can write the category description for each pile of cards).
8. When the user has completed the task, document what cards were placed in each group.
9. When the participant leaves the session, shuffle the cards again before the next participant arrives.
10. Follow Steps 6–9 for the remaining participants.

Analyzing the card sorting can be easy or complex, depending on the results of the card sorting. First look at the groups created by the participants. What cards were sorted into groups? How many of these cards were sorted into the same group? Where were the differences? If there is quite a bit of overlap in the card sorting, you can assume the card placement is representative across users and thus use this taxonomy for your website. However, further analysis may be needed if participants' card groups are more varied. In this case, creating a matrix of results or percentages of cards in each section may be necessary. You may even find that testing more participants will be necessary. Ma (2010) and Sauro (2012) both have excellent resources on how to analyze card sorting that you should consult.

Kate ran a card-sorting activity to categorize the methods presented in this chapter. To do this, she created a card for each method. On the front was the name of the method; on the back was a definition for the method. Definitions were given so users with less research experience could read more information about each method before deciding how or where they would group the method. Figure 4.19 shows the results of the card-sorting sessions.

4.6.5 User Profiles and Personas

Your goal, after you've carried out user research, is to create user profiles or personas. Which one do you create? You could actually do both because they both serve different purposes.

A user profile is a set of defining characteristics, based on your research, that represents a particular group of users. You may find that your product has more than one

FIGURE 4.19   Card sorting groups. (Courtesy of Dr. Kate Crane.)

group of users. Each group, in other words, has its own defining set of characteristics that make it a little different from other groups. You can choose to rank these groups in order of importance and then concentrate primarily on developing your product with the most important group's needs and wants in mind. Conversely, you can choose a cross-section of important characteristics from all the groups and create a meta-group, one representative of all the groups that you then use to recruit users to participate in the design process. Which should you choose? It depends. Obviously, if you choose to design for only one group, you risk ignoring the needs and wants of other groups. But it could be the product you're developing, such as the law school website example at the very beginning of the chapter, has one user profile, when broadly defined, that makes up a significant percentage of the potential population that will use your product. It could be that your product only serves, and will only serve, a more focused population. It could also be that you choose a group that you know has such important needs and wants that if your product can meet those, then you will likely address, at the same time, the issues of other groups.

Personas are fictional characters created to represent an actual user. There are, according to Cooper, three different types of personas: marketing, proto-personas, and design (Ilama 2015). Our focus is on design personas. According to Ilama, design personas "tell a story and describe why people do what they do in attempt to help everyone involved in designing and building a product or service understand, relate to, and remember the end user throughout the entire product development process" (sec. 2, par. 4).

We create personas to serve as active reminders of our users during the design process. Whenever we make a design choice, we look at a persona and ask ourselves, would this person need or want it and would it achieve this person's goals? Additionally, should a design team have disagreements, such as whether a new feature should be added, the persona helps solve the dilemma, keeping the process focused on tangible user goals as expressed through the persona's description (Calabria 2004).

This persona description, because it is intended to be an embodiment of a perceived user of a product, is rich in detail. When we create a persona, we try to create the closest representation of a real-world user who will want or need to use our product. Ilama (2015) notes that a good design persona (Figure 4.19), among other things, should be driven by research and contain information about user context, behaviors, attitudes, challenges, and pain points (problems) as well as goals and motivations.

4.7 EYEGUIDE FOCUS CASE STUDY

Now that you've learned many methods for researching users, let's take a look at an example of how all this work can be brought together to generate a user profile and persona.

From this point on, we will rely extensively on real research, including visual documentation, that EyeGuide, Brian's eye-tracking and control technology company, carried out to develop what is an actual product in use around the world: EyeGuide Focus. Focus is a 10-second concussion detection system that relies on wearable eye-tracking hardware in conjunction with software that runs on an iPad to administer reliable, fast concussion tests for athletes who may have suffered mild traumatic brain injury or concussion while playing sports, such as football or soccer. Many other concussion-detection systems take longer to work, cannot be used during actual play, or don't rely on quantitative measurements to make determinations about concussion.

Because of the obvious need for fast, reliable concussion detection, given the increasing rate of concussions among athletes, especially in youth sports in the United States, EyeGuide pivoted away from its previous product offerings to embark on building out and deploying Focus to the market, beginning in late 2014. Such a pivot was challenging, requiring the company to learn about an entirely new user population and its needs and wants. EyeGuide didn't want to make the same mistake it experienced when attempting to offer up Assist (see Chapter 1) as a low-cost mouse replacement for users with limited or no hand functionality. So, from the outset, it employed a UCD process. In fact, the same RABBIT process described in this book is what EyeGuide used to create and deliver Focus to users.

Figure 4.20 is the storyboard that EyeGuide developers sketched out early in the researching users phase.

You can see the storyboard was EyeGuide's early attempt to understand how Focus users, namely athletic trainers, would need the product to perform given their use environments. Most athletic trainers have to take care of dozens, if not hundreds, of athletes. They work in dynamic conditions—different sports; different fields of play; and busy, noisy locker and training rooms. EyeGuide used the storyboard to project how Focus would need to be designed to be integrated effectively for athletic trainers.

User research didn't stop there. Triangulation was key. In addition to interviewing a number of athletic trainers as well as discovering and reading articles and other publications written about athletic trainers and their experiences with concussion

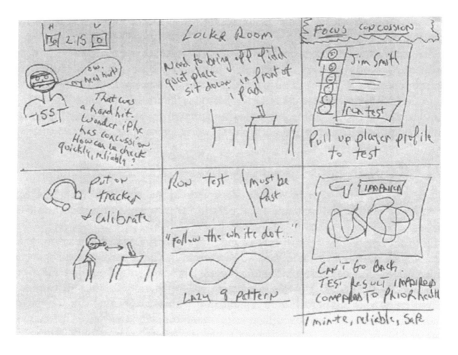

FIGURE 4.20   Focus storyboard. (Courtesy of EyeGuide.)

screening, EyeGuide developed a use case (Figure 4.21) and went on location to conduct site visits with select athletic trainers.

The site visit, as noted earlier, is a comprehensive research tool for UCD because it comprises fully the data it collects in the *see-say-do triangle*. EyeGuide learned more about athletic trainers' environments. Athletic trainers walked the EyeGuide developers through the existing processes they used for detecting concussion,

Use case 1: concussion detection

• Pack (nitrogen + base board) are plugged into AC power

• User wears the device while sitting down

• User looks at a device screen (can be any device, i.e., iPad, desktop/laptop, tablet) and watches dots or other reference points or objects to collect time, fixations, saccades, or other visualization data

• Both the eye and scene cameras are used

• Total test time is roughly 5 minutes

• Multiple users may be tested back to back as part of baseline testing process, but when testing for abnormalities in users testing will be on-demand and so less frequent

FIGURE 4.21   Focus use case. (Courtesy of EyeGuide.)

verbalizing their thoughts about the strengths and weaknesses of what they did. Finally, EyeGuide developers observed how athletic trainers interacted with athletes.

EyeGuide was able to assume at the beginning that athletic trainers would be the primary audience for Focus, and that, in a sense, made the development of a user profile and persona easier. A problem with Assist that would have even challenged a UCD approach, had it been implemented, was that there are so many different types of possible users and thus profiles (personas) of assistive technology, many of which have competing needs and wants or would use the product in different environments. For EyeGuide's purposes, Focus was designed initially to be a sideline concussion detection tool. Other users, such as health care practitioners at clinics or hospitals, although of interest for future development, did not come into play for this first production version of Focus. Thus, they were not considered.

After accumulating a great deal of data about athletic trainers, EyeGuide carried out affinity diagramming to synthesize everything into a manageable collection of characteristics, which the team collectively agreed upon, that became the basis of the user persona representing the target audience for Focus. It was this persona that would also be used in subsequent phases of the RABBIT process to recruit users for trying out prototypes during the development process. Figure 4.22 shows the user persona that was created to represent EyeGuide Focus's primary user.

As development began, EyeGuide created other user personas. Certainly, you're not limited in the number of personas you generate. However, if you find yourself making so many that it complicates your ability to make decisions about your product's design, your team may need to discuss how best to combine personas and prioritize particular personas over other ones based on company goals, time, budget, and other factors. You might also choose to eliminate from the process those personas

Kimber Johnson

Female
31 years old

Occupation: certified athletic trainer (CAT)

Employer: Large (2,000 + enrollment), public high school in Texas

Education:
• BA, exercise and sports science
• Masters in athletic training (MAT)

Salary: $50,000

Technical profile: comfortable with new technology—often required to adopt, for example, new health care tools for athletic training work. However, has limited time especially during the season for learning new tool/process that might be more time-intensive.

• "I don't want to replace any of the tools I currently use for concussion detection. But I do want to make detection and recovery monitoring better, faster."

• "It would be great if I could use the tool to detect a concussion right away."

• "It would be better if I could monitor recovery from concussion. For all of our training, there is just too much uncertainty about knowing when an athlete is really ready to return to play after a concussion."

Kimber has been a certified athletic trainer (CAT) for 5 years at this high school. Before that she interned as an athletic trainer trainee for an NCAA football bowl series (FBS) university. She reads new literature on concussion research as well as other research related to athletic training and injury, rehabilitation care.

She is comfortable with existing concussion detection but also thinks there are drawbacks, such as how some take to implement, or the possibility athletes can cheat the scoring.

Kimber is actively involved in supporting athletic trainers around the area where she works. She believes athletic trainers at small schools could use a more reliable, easier to use system because they lack resources, i.e., money and medical support staff, to help them make decisions.

FIGURE 4.22   User persona for EyeGuide focus. (Courtesy of EyeGuide.)

that don't represent your product's primary users. What's a good rule of thumb for the right number? Barnum (2011) says "more than 2 but less than 12" (98). EyeGuide elected to create just one. Given time constraints, forced into having to build and deploy an early stage or alpha version of Focus as soon as possible, the company's developers felt like focusing on the needs and wants (as well as situation) of just one user helped them narrow the scope of Focus so that it could be managed effectively enough to deliver something that worked as well as something that could be finished for field trials as soon as possible.

This look at researching users was in depth. Although it easily serves as the largest chapter in our book, it is worth the concentrated examination. Everything we do in UCD starts and ends with users. You can change, and you should, your understanding of users as you gain more knowledge about them. But before you ever design anything for them to use, you need to start with a firm, foundational understanding of who they are, what they want and need, and where they work.

Now it's on to the next phase of RABBIT: assessing the situation of use.

TAKEAWAYS

1. Always triangulate data from the client and other sources to get the most accurate picture of the user and potential use scenario.
2. Don't make assumptions about who your users are. To understand them, you must draw from a variety of methods, including those that involve direct interaction with users where they work and live.
3. Choose user research methods that best suit the purpose of your design and user inquiry. A diary study may not be the best method if you only have two days to study your users, and card sorting may not be beneficial if you are observing a doctor input patient information during a 10-minute lunch break.
4. Observation methods are one of the most useful tools to understand user workflow, tasks, and environment. Observations help the designer understand the use scenario and provides a realistic view of the conditions in which the design will be used.
5. A user analysis, task analysis, and environment analysis are critical parts of a successful site visit.
6. User self-consciousness, which can be amplified when being observed or frequently asked questions, can cause users to complete tasks differently than they normally would. Try to minimize this by using a variety of methods.
7. Be mindful of confirmation bias or using observations to confirm existing beliefs. Using multiple observers and/or following up observations with verbal data (such as informal interviews) can be used to cross check observational data.
8. Users make emotional decisions about products. Emotion heuristics, an enjoyability survey, and product reaction cards can help you understand users' emotional views of a product.

9. What users say they want may not correlate with what they actually need or desire in a design.
10. If you do not research a truly representative user, your design can go astray. Make sure you observe the right (representative) user in the right (representative) use scenario.
11. Your understanding of users will evolve as you gain more knowledge about them.

DISCUSSION QUESTIONS

1. Why is it helpful to gather data about users from multiple sources? If a client hires you to design a product with a specific user in mind, how might you explain that you'd like to gather additional information about potential users without offending the client?
2. When might broader analytic data be helpful in understanding user behavior? When might you need more focused, in-depth data?
3. What are the benefits and drawbacks of observation and self-reporting? When might you use each method?
4. What are the benefits and drawbacks of site visits, shadowing, and fly on the wall? When might you use each method?
5. How can you avoid confirmation bias? What if you are working for a client who gives you a specific list of user flows or problems and wants you to immediately figure out how to fix them?
6. Why does user emotion matter in the UCD process? Can't you just look at purchasing data and user demographics?
7. How does creating a user flow, task flow, or journey map help in the design process?
8. Why is a user profile or a persona helpful?

REFERENCES

Barnum, C. *Usability Testing Essentials: Ready, Set...Test.* Burlington: Morgan Kaufmann, 2011.

Benedek, J., and T. Miner. "Microsoft Product Reaction Cards." *Microsoft.* 2002, accessed February 20, 2016. www.microsoft.com/usability/UEPostings/ProductReactionCards .doc.

Calabria, T. "An Introduction to Personas and How to Create Them." *Step Two.* March 2, 2004, accessed February 21, 2016. http://www.steptwo.com.au/papers/kmc_personas/.

De Lera, E., and M. Garreta-Domingo. "Ten Emotion Heuristics: Guidelines for Assessing the User's Affective Dimension Easily and Cost-Effectively." Proceedings of the 21st BCS HCI Group Conference, Lancaster University, United Kingdom, September 3–7, 2007.

Flom, J. "The Value of Customer Journey Maps: A UX Designer's Persona Journey." *UXmatters.* September 7, 2011, accessed February 21, 2016. http://www.uxmatters.com /mt/archives/2011/09/the-value-of-customer-journey-maps-a-ux-designers-personal -journey.php.

Hackos, J., and J. Redish. *User and Task Analysis for Interface Design.* New York: John Wiley & Sons, 1998.

Helen Hamlyn Centre for Design. "Designing with People." n.d., accessed February 21, 2016. http://designingwithpeople.rca.ac.uk/.

Holtzblatt, K., and H. Beyer. *Contextual Design Evolved*. San Rafael: Morgan & Claypool, 2015.

Hughes, M., and G. Hayhoe. *A Research Primer for Technical Communication: Methods, Exemplars, and Analyses*. New York: Routledge, 2008.

Ilama, E. "Creating Personas." *UX Booth*. June 9, 2015, accessed February 21, 2016. http://www.uxbooth.com/articles/creating-personas/.

Information and Design. "Affinity Diagramming." 2012, accessed February 21, 2016. http://infodesign.com.au/usabilityresources/affinitydiagramming/.

Kenworthy, E. "Use Case Modeling: Capturing User Requirements." December 1997, accessed February 21, 2016. http://www.zoo.co.uk/~z0001039/PracGuides/pg_use_cases.htm.

Kujala, S., and M. Kuappinen. "Identifying and Selecting Users for User-Centered Design." *Proceedings of The Third Nordic Conference On Human-Computer Interaction*, Tampere, Finland, October 23–27, 2004: 297–303.

Kuniavsky, M. *Observing the User Experience: A Practitioner's Guide to User Research*. San Francisco: Morgan Kaufmann, 2003.

Ma, S. "Dancing with the Cards: Quick-And-Dirty Analysis of Card-Sorting Data." *UXmatters*. September 20, 2010, accessed February 21, 2016. http://www.uxmatters.com/mt/archives/2010/09/dancing-with-the-cards-quick-and-dirty-analysis-of-card-sorting-data.php.

Miller, D. Listserv contribution used with author's permission (1997), quoted in Hackos, J., and J. Redish, *User and Task Analysis for Interface Design* (New York: John Wiley & Sons, 1998), 69.

Recknagel, J. "Shadowing." *Design Research Techniques*. n.d., accessed February 22, 2016. http://designresearchtechniques.com/casestudies/shadowing/.

Sauro, J. "10 Things to Know About Card Sorting." *Measuring U*. April 3, 2012, accessed February 21, 2016. http://www.measuringu.com/blog/card-sorting.php.

Seels, B., and Z. Glasgow. *Exercises in Instructional Design*. Columbus: Merrill, 1990.

Smith, A. "U.S. Smartphone Use in 2015." *Pew Research Center*. April 1, 2015, accessed February 20, 2016. http://www.pewinternet.org/2015/04/01/us-smartphone-use-in 2015/.

Sparandara, M., L. Chu, and B. Enge. "Uncovering Context with Mobile Diary Studies." *Punchcut*. 2015, accessed February 20, 2016. http://punchcut.com/perspectives/uncovering-context-mobile-diary-studies/.

Usability.gov. "How To & Tools: Use Cases." *Usability.gov*. n.d., accessed February 21, 2016. http://www.usability.gov/how-to-and-tools/methods/use-cases.html.

Usability Net. "Focus Groups." *Usability Net*. 2006, accessed February 20, 2016. http://usability net.org/tools/focusgroups.htm.

SAMPLE SITE VISIT

Lena Thompson, earth science teacher at a midsized public high school in suburban Chicago, Illinois
Maggie Waldron
October 7, 2015

INTRODUCTION

On Wednesday, September 30, 2015, I visited a midsized public high school (MPHS) in the suburbs of Chicago, IL, to shadow and interview Lena (pseudonym), who is a science teacher and chair of the science department. My visited lasted 6 hours, beginning at 7:30 am and ending at 1:30 pm. Lena and I have worked together in the past to develop science curriculum and lead teacher professional development.

USER ANALYSIS

I reached out to Lena for this site visit because she and I have worked together in the past and we have a good working relationship. Furthermore, her school is not implementing the STEM education program I run, so I could expect her to have many "novice-level" questions about it even though she is an expert educator. Even if it didn't come up in conversation, I wanted to shadow someone who didn't have full knowledge of the program. Since the goal of my UX project this semester is to redesign the program's website, I thought it would be useful to shadow a potential, rather than current, user of the program. As Lena has both teaching and administrative roles, I hoped to learn how each of those sets of responsibilities might influence her experience using the future website.

METHODS

This site visit incorporated user shadowing followed by a semi-structured interview. I shadowed Lena during morning rounds of teacher planning time and sat at the back of her classroom while she taught her earth science course. When she's not teaching her own class, Lena spends much of her day conducting observations of other teachers in her department. Some of these observations are informal and others serve evaluation purposes and are therefore confidential. As a result, there were parts of her day in which I could not participate. During those times, I observed other teachers in the building. While those observations were tremendously enjoyable—MPHS has a terrific science department—I will not include them in this site visit because they do not directly pertain to Lena's daily experiences and responsibilities. Before leaving, I spent nearly one hour interviewing Lena to get a sense of what she had been doing when I wasn't able to shadow her. I'll try to fill in the blanks of her day using the information she provided.

USER PROFILE

Lena has taught at MPHS for 13 years. Before that, she taught at a nearby high school (in the same district) for 9 years. She has a M.A. in earth and environmental sciences. She always enjoyed being a Teaching Assistant (TA) in grad school and earned a teaching certificate afterwards; she has been teaching ever since.

I asked Lena to describe her responsibilities at school and beyond. She listed several roles that occur both during and after the school day. At school, she is a science department chair, co-chair of the technology integration program at the school, teacher, and a presenter to other school committees (i.e., the parent advisory board). As science department chair, she is "responsible for establishing a vision for the science department that includes day-to-day operations and safety to planning 4–5 years down the road for what instruction will look like." As co-chair of the technology integration program, Lena mentioned that she is responsible for organizing trainings. By way of explanation, she offered that science teachers are often the early adopters of new technologies in the school. (I observed that all of the science teachers I observed integrated the use of Google Drive into their class in some way, whether to deliver assignments/ worksheets to students via their class folders, outline the day's agenda, or in some other productive way.)

Lena's collection of presenter badges from numerous
workshops and other professional development.

Outside of her official roles at MPHS, Lena does a considerable amount of consulting work with several professional and educational organizations. She has been involved with devising state- and federal-level strategies for implementing the Next Generation Science Standards (NGSS), and she also helps evaluate new science curricula. All of this work, she reflected, does influence her vision for the science department at MPHS.

When I asked her about what she does for her own professional development (PD), Lena's immediate response was, "That's a hard one." She described herself as a "Renaissance woman" who is mostly self-taught. I asked if it would be fair to describe her as being on the leading edge of innovative science education, and she agreed. She said she doesn't take classes but identifies needs and learns through the process of assembling teams to teach each other through leading teacher professional development workshops and other similar experiences. Lena reflected that it would be easier to describe her PD as a massive concept map of her network of connections, and she pointed to the collection of "Presenter" badges in her office that she has accumulated over the years (see photo).

ENVIRONMENTAL ANALYSIS

MPHS is a public school located in a suburb of Chicago, Illinois. In 2013–2014, the school's total enrollment was 1,923 students (the most recent year for which I could find data). The school's online profile describes the surrounding community as residential with some industry, offices, and a diversity of low- to middle-income homes. In 2013–2014, the school's population included 47.2% low-income students. Students at MPHS speak more than 50 different languages; 75% of students live in homes where English is not the primary language. In 2013–2014, the student body was approximately 40% white, 30% Asian, 20% Hispanic, and 8% black.

Visitors to the school are greeted by a security guard who ensures each person signs the visitor log and receives a visitor's badge. Two long hallways and the main office stem from the entryway. After the security guard handed me my badge, he called Lena's office phone to let her know I had arrived, then sent me down the right-hand hallway toward her office to meet up with her.

Lena's office is on the first floor. It contains a work table that seats four to six people and a large L-shaped desk where she has her computer and workstation set up. She noted that having that work table was indispensable. The room has been painted a bright green, and every time I have been in there (for this visit and previous ones), she has been playing music in the background. On a countertop along one wall, she has on display a collection of several different types of rocks, and there are a couple photographs and a small painting hanging on the walls.

Over the course of the day, Lena and I spent time in several different science classrooms around the school. In general, all of the classrooms were designed in the same way (see photos), with a group of small work tables set up in the front of the room and permanent lab benches installed in the back of the room. It seemed that students sat at the work tables during lectures, presentations, and any other work that did not involve running experiments or other lab activities. Presumably, students migrate to the rear of the classroom when they are doing experiments. There were some variations among the classrooms depending on which subjects they housed. The earth science classroom had field equipment lying around: things like nets, sieves, plastic tubs, etc. The biology classroom contained fish tanks, plants under a grow light, and topically relevant posters (i.e., describing the characteristics of living things). In the chemistry classroom, each lab bench had assorted lab supplies, including various glass containers filled with various reagents and solutions, neatly organized in the center. The physics classroom was notably less decorated than the others; its main decorations were a collection of words related to the practice of science, including: "predict, anticipate, confirm, summarize," etc.

Lab benches in the rear of Lena's earth science classroom.

TASK ANALYSIS

Immediately after I arrived, Lena set off to the first set of observations of her day. I was allowed to shadow her during this activity as these observations were informal and would not be used for evaluation purposes. On this particular day, the hour before school started was dedicated to "professional learning group" (PLG) time, during which teachers work with their subject teams to develop uniform assessments and other curricular goals that ensure instruction is relatively similar across their classes. The first PLG group we visited was the biology team, which was comprised of five female and two male teachers.

Lena briefly introduced me to them and explained that I was here to observe, and then the teachers quickly returned to their conversations. Lena walked over to one group of teachers and listened to them describe what they were working on. The other teachers were working either in pairs or individually; one had headphones in, which she said was to help her focus. Some teachers were using pen and paper to work; others were working on laptops. After about 10 minutes, Lena cued me that we would head to the next PLG group.

Next, we stopped in to observe the chemistry teachers who were working in another classroom down the hall. In this group, there were three female and four male teachers. They were animatedly discussing a grading rubric for a pre-lab assignment in which students were supposed to set up their data collection table. One teacher sat by a laptop connected to a projector, which was displaying a Google Sheet at the front of the room. Lena sat at the edge of the group and listened for the most part, nodding occasionally. She chimed in only to agree with one of the teachers about which grades should indicate that a student is below/meeting/exceeding standards, and when they should be referred to academic support.

After another 10 minutes, we walked to the next classroom to visit with the PLG of physics teachers. There were only three teachers in this group: two male and one female. From their conversation I learned that another male teacher was out sick; it also seemed that he was the lead teacher of this group and the one most familiar with their online grading system, which the three teachers present seemed to be struggling with. Lena patiently watched while one of the teachers showed her what he was trying to accomplish in the grading system. She later commented that she was aware that these three are less confident with technology than the teacher who was absent, and she wanted to let them figure this tool out on their own.

After about eight minutes, we headed to the classroom next door to join the earth science PLG. The three teachers present here were all male; Lena is the fourth teacher in this group and the only female earth science teacher at this school. One teacher stood at the front of the room and controlled a laptop that was projecting the online grading system and a Google Sheet document on the front whiteboards. The other two teachers present had their laptops open and were editing the same Google document as the conversation progressed. The discussion in this room also centered on the online grading system, which was giving this group trouble as well. They were having trouble seeing each other's assessments, so Lena logged in to her account to pull up some of the master files.

Immediately following this last PLG observation, Lena had to do her teacher evaluation so I spent the next class period observing two different biology classes. Afterward, I joined Lena again for her earth science class, which ran from 9:50 am to 11:00 am. She told me in advance that on Wednesdays (including this day) her students worked on their group projects, which were described to me by another teacher as time to explore earth science careers.

As soon as the bell rang, Lena directed her students, "Ladies and Gentlemen, please grab your Chromebooks," while shutting the classroom door. (At this school, all students have Chromebooks that they carry to all of their classes; these are utilized to differing extents by various teachers in the school.) She then notified them of a new document in their Google Drive that she had recently added to their class folders. She instructed them to read it, then close their laptops half way so she would know they were ready for the next step.

On this day, there were 23 students in the class. All of them were seated before the bell rang except for one girl who arrived several minutes late. Based on the way Lena greeted her at the door, it seemed that this was a common occurrence for her, in particular. Nearly all of the students were very attentive while Lena was speaking to them.

Lena then introduced the topic of Student of the Month to her class. She asked them to discuss with their neighbors what qualities a student would need to have to qualify for this recognition. As students were volunteering responses, Lena circulated around the room, holding the certificate but not acknowledging yet who was to be the recipient. Student responses included, "trying your hardest, being a good role model, helping others, contributing in class." Before announcing the Student of the Month, Lena acknowledged that she had picked this month's winner, and the class asked if in the future they can have a discussion about who should get it and why; Lena agreed to this but asked for veto power (to which they conceded).

Next, Lena told the class they would resume work on their group projects. "This is not a research project," she reminded them. "This is a call to action." She passed around an article on best practices for using social media to promote small businesses and encouraged them to think about how they might be able to use social media to create calls to action for their projects. Anticipating that some students would be surprised that their teacher was encouraging them to use social media at school, Lena said, "We use the right tool at the right time. If it's the right tool, why wouldn't we use it?" She instructed students to read in partners, reminding them they weren't "studying for a test, we are being purposeful readers." In other words, she did not want them to memorize everything and try to apply all the tips to their project; she wanted them to decide which one(s) would be the best fit. After students finished reading, she told them to gather with their project groups and draft their weekly summary of their progress. Lena's emphasis to her students was always on thoughtfulness and practicality. She reminded them that their project proposals were really just "drafts," that they were works in progress and could be revised continually as their projects took shape. "Use tools that will work for you," she emphasized. "Don't be conventional if it will limit what you can do."

After Lena dismissed her class, she headed back to her office to take a conference call and suggested that I observe another class while she did so. I did not ascertain whether she thought it wouldn't be important for me to listen to that call or whether she would prefer I wasn't present, but I didn't want to press the issue and figured I could ask her about it in our interview later. I rejoined Lena at noon in her office, when her conference call had concluded and she had indicated she had some free time to answer my questions.

REFLECTIONS

My goal for the final interview session was to ask Lena a series of questions to help me understand her daily routine, reflect on various parts of her day (the parts I had observed as well as those I could not), the challenges she faces at work, and what her goals are in each of her many roles.

I started by asking about what she is thinking about when she does her rounds among the PLGs. She described one of the goals of the PLGs as attempting to find more resources for students because "they want them to connect with science." Additionally, she said that her team wants to be sure that when they give extra [academic] support to students, it's the right kind of support and provided at the right time (i.e., not too late to be effective). She mentioned that the PLGs have facilitated the spread of certain practices and useful tools throughout the department in advantageous ways.

Then, I asked Lena to describe for me what it's like for her to do a teacher evaluation since I wasn't able to observe that part of her day. She showed me their evaluation rubric, which she described as "very prescriptive." It is based on the Danielson framework for teacher evaluation. She stated that her evaluations of teachers are comprehensive and described the series of documents, interviews, informal conversations, formal observations, and other assorted data that she compiles all year long for each teacher. She made a point here to connect good teaching to the STEM program I work for: to be evaluated well according to Danielson, student engagement is absolutely necessary. (Promoting youth engagement is a central tenet of the program's model.) According to Lena, tenured teachers have formal observations every other year; non-tenured teachers are evaluated every year. All teachers are observed informally on a quarterly basis.

I then asked Lena to describe the schedule of her typical day. She told me that she arrives at school at 6 am and that the rest of the teachers usually arrive at 7. She called them "her teachers," which I noted in my observation log. Often, mornings include "dealing with subs and sub plans" by which I assumed she meant securing substitute teachers and ensuring they had adequate plans for the day. She noted that her teachers like to stick around school at the end of the day, and that they are all there, working and/or hanging out, until about 5 pm. She implied that because her teachers are there, she would be too, because time spent with them is valuable.

Right away, Lena pointed out that her daily responsibilities change over the course of the year. From the beginning of school until November, her non-teaching hours are occupied by "getting the day going" (i.e., sub plans) and teacher observations. Mid-year, she is organizing teacher observation plans (for evaluation) and writing recommendations. During the early winter, it's "all about scheduling": She and her teachers spend a lot of time reviewing students' transcripts to ensure they're enrolled in the right classes. The early spring, she simply described as the "craziness of AP testing." In general, she said, there are "way too many" administrative meetings. She tries to maximize her time working with teachers "off the books." This includes "having conversations with them about what's going on"; this is specifically *not* feedback but "just supporting them," and it's where she feels she needs to spend most of her time. She said she prefers to put off writing reports until the end of the day because she would rather be in the classroom to talk with teachers. She has to work hard to make time for this, as she feels she could easily spend three to four hours per day writing and responding to emails.

Since I value Lena's teaching experience and perspective on STEM (science, technology, engineering, & math) education, I then asked her what she looks for in new STEM curriculum.

The biggest problem [with new curriculum or other initiatives], she said, is that teachers already feel like they have too much on their plates. She emphasized that it was critical to prove the value of any new effort to help teachers feel like they can move other tasks out of the way to make room for something new. It "needs to feel like it's an all-encompassing opportunity that could meet the needs of all students," she said. Another potential concern she would have introducing new STEM curriculum is that high schools are "hugely territorial, a land-grab for students." Teachers, she said, can become worried about whether they'll lose enrollment for their own classes if something new comes along. Furthermore, in her opinion, high school administrators perceive their teachers as highly specialized and may assume or expect them to bring in new resources or opportunities for their students on their own. In contrast, Lena's opinion is that both the territoriality and administrative expectations might be less in the middle schools although she was clear that she has not taught in a middle school so was just making a guess.

Lena and I discussed that the new website my team is working on would need to have a clear value proposition that's almost immediately accessible to a new visitor. It would need to distinguish itself as not just "something else [for teachers] to do." Additionally, the new website would need to help the program stand out not just as part of the STEM education trend but as a nimble program that adapts easily and well to a variety of school settings and needs.

My last question to Lena was "What makes a STEM program stand out to you?" Her responses were extremely useful: First of all, the program must be relevant to what's happening in STEM education in the near term. When I asked her to tell me what she thought would be relevant in the near term, she replied that programs need to offer ways to "open kids' eyes," particularly those kids "who tend to be under-dreamed." These programs cannot feel contrived and must not make kids feel like they're being talked down to. In addition, they must tie in with a school's goals and vision for their students.

On a more practical note, she also said that teachers want to know that there's always "someone on the other end of the line," willing to answer questions and provide support throughout the year.

CONCLUSION

Through my observations of Lena and our subsequent conversation, I gathered many useful insights during this visit that helped me to generate a list of the kinds of information educators like Lena might want to find during their initial visit to our future website. Some of the key takeaways for me include the following:

- The way Lena modeled thoughtful research and problem-solving skills demonstrated the value of highlighting how the STEM program can help students develop "21st-century skills."
- Lena's practical concerns about teachers' workloads means that any new design of the website should spell out how easy it is to start up, implement, and maintain the program.
- Her attention to the importance of student engagement as a core goal of her science department, and of a quality educational model in general, suggests that the website needs to emphasize this as a central goal of our program; we should also include stories or images that drive this point home.
- Lena also mentioned the importance of any STEM program being all-inclusive and able to reach "under-dreamed" youth in a meaningful, authentic manner. These are demonstrated strengths of our program (as observed by staff and through teacher reflections), and they should be clearly communicated through the website as well.

SAMPLE SURVEY TO DETERMINE 2311 REPRESENTATIVE POPULATION

This survey is part of the study "The Usability of the Course Syllabus: Testing Syllabi Modality and Comprehension" conducted by Dr. Kate Crane. This survey's purpose is to collect demographic and behavioral data about course syllabi to assist in selecting a representative sample of English 2311 students for usability testing. This study has been approved by TTU's Institutional Review Board for Human Research. The data collected will be anonymous and used only for research purposes by the investigator. If at any time you do not wish to resume this survey, you may stop without penalty. Should you have any questions about this research, please feel free to contact Kate Crane at kate.crane@ttu.edu.

1. With what gender do you identify?
   a. Female
   b. Male
   c. Other

2. Age (multiple choice)
   a. Under 18
   b. 18–23
   c. 24–30
   d. 31–39
   e. 40–49
   f. 50+
3. Year in college (multiple choice)
   a. 1st year (freshman)
   b. 2nd year (sophomore)
   c. 3rd year (junior)
   d. 4th year (senior)
   e. 5th year +
4. What is your G.P.A.? (open-ended)
5. In what TTU college are you enrolled? (multiple choice)
   a. College of Agricultural Sciences & Natural Resources
   b. College of Architecture
   c. College of Arts and Sciences
   d. Rawls College of Business Administration
   e. College of Education
   f. College of Engineering
   g. College of Media and Communications
   h. Have not enrolled in a college yet
6. What is your major area of study? (open ended)
7. What is your primary reason for taking English 2311? (multiple choice)
   a. Required course for my area of study
   b. Fulfill writing requirement
   c. Advisor recommended the course
   d. Interested in technical writing
   e. Other (with room for text)
8. What grade do you expect to receive in 2311? (multiple choice)
   a. A
   b. B
   c. C
   d. D
   e. F
9. Where do you look for course information most often? (multiple choice)
   a. Moodle
   b. The syllabus
   c. I ask classmates
   d. I ask my instructor
   e. I ask questions in class
   f. Other (with room for text)

10. How do you prefer to receive course syllabi?
   a. In print
   b. Electronic file (PDF or Word file)
   c. Web page
   d. No preference
11. How was your course syllabus delivered? (multiple choice)
   a. The instructor gave me a print copy only.
   b. The syllabus was posted on our Moodle course as a PDF or Word file only.
   c. The syllabus was given as a print copy and posted on our Moodle course as a PDF.
   d. The syllabus was delivered as a web page only.
   e. The syllabus was delivered as a web page and linked to our Moodle course.
   f. Other (with room for text)
12. How often do you refer to the course syllabus throughout the term? (multiple choice)
   a. Never
   b. Rarely (1–3 times)
   c. Occasionally (4–9 times)
   d. Frequently (10–15 times)
   e. At least once a week (16+ times)
13. How do access your course syllabus? (multiple choice)
   a. Online from a Moodle file or link
   b. From a print copy I received from my instructor
   c. From a print copy I printed from Moodle
   d. Electronically from a file I saved from Moodle
   e. Other (with room to add text)
14. What information do you consult the syllabus for? (ranking question)
   a. Course outcomes
   b. Course assignments
   c. Course grading procedures
   d. Course policies
   e. Instructor information, such as contact information and office hours
   f. University policies
   g. Other (with room to add text)
15. In 2311, what types of resources do you read online? (check all that apply)
   a. Course assignment descriptions
   b. Course textbook
   c. Supplemental readings
   d. Assignment feedback from my instructor
   e. Research resources (journal articles, newspaper articles, books, blogs, etc., for my own research projects in the class).

16. What device(s) do you use most often when accessing course materials? (multiple choice)
    a. Laptop computer
    b. Desktop computer
    c. Netbook
    d. Tablet
    e. Smartphone
    f. Other (room for text)

5 Assess the Situation

In the previous chapter, we focused on learning methods to help you know who the users are for the product you will design. Assuming you've compiled a decent (although never complete) understanding of your users' motivations, experiences, desires, and capabilities, now is the time to assess the situation that surrounds not only where your users will use your product, but also the situation that influences how you go about creating that product.

Your product's users live in the real world and will use this product with their own expectations and experiences based on what they've learned, done, or do in the real world. You also will more than likely make a product that has competition, or if there is no competition, then it is similar in materials or functionality to other products people use or have knowledge about. Finally, you're going to have deadlines, goals, budget constraints, and other factors that contribute to how you go about designing a product for its intended users. Particular analytical methods, which we list below and then explain in greater detail, will help you effectively assess both the situation that surrounds your design process and the situation in which users will use your designed and delivered product.

Namely, in this chapter, we define and explain how to do the following:

- **Functional analysis:** What must the product do?
- **Environmental analysis:** Where will the product be used, and how should we account for that so the product integrates effectively into the environment if not actually enhances the environment for the product's users?
- **Organizational analysis:** What are your organization's goals for the product? What resources will you be provided? What's the budget? What limitations or other factors will impact how you design the product?
- **Competitor analysis:** What other products are like yours? What about yours will be different? What white space can you occupy?
- **Materials analysis:** What will go into the construction of your product? How will this impact its design and how users use it? What products may users have to interact with what you've built for them that you need to consider?
- **Content analysis:** What content has to go into your product to satisfy users' needs and wants? What can be left out?

At the conclusion of the chapter, which offers up examples again of the Focus concussion detection system design project, you should have all that you need to move to the next stage of RABBIT.

5.1 FUNCTIONAL ANALYSIS

You want to detail the functional requirements of the product as part of your assessment of the situation. To do this, start with everything you and, if applicable, your organization wants the product to do. Look as well at what other products do.

Certainly, too, take advantage of any user preference data you have, such as interviews, surveys, and focus groups, in which users have offered up their opinions about what they want the product or products like it to do. *After all this, however, it is absolutely crucial that you determine the actual required functionality of the product.*

People love to give you a wish list. You certainly have design wishes, as does your organization. But the functional requirements you build into the product should be those that allow users to successfully accomplish real tasks. One idea is to develop a list of features and, next to each item, explain what it does. Go back to your task and user analyses and determine if the feature's functionality meets users' needs so they can carry out tasks. Those features that don't accomplish anything for the users should be downgraded in priority or removed altogether.

Putting together this features list also helps determine if you have redundant features—if so, merge them together or eliminate one. Once you've narrowed down your features list, walk through each feature as though you were actually using it to do what it's supposed to do. Better yet, have users do this. If users can use the feature and if they need it, you've got an important feature you need to keep.

Doing this might also help you be prepared for any functionality that might cause problems for users. Sometimes, one feature's functionality can impact another feature. By having users walk through each feature's functionality, you'll be able to spot a feature that doesn't make sense or that causes problems for another feature. You can, if the feature isn't necessary, eliminate it, but if it is necessary, you can already think of ways you'll need to design the product to have cohesive, necessary functionality.

You can see in Box 5.1 the efforts EyeGuide developers made to understand the functions that had to be in the Focus concussion detection system, not only to meet user needs, but also to be competitive against competitor offerings.

The format for what Focus presented was self-designed, based on previous conversations within the team and research done on the market. Your functional requirements need not be as rich nor follow the same format. Vijak Sankar (2013) notes that typically "functional requirements provide the high level description of how a system or product should function from the end user's perspective. It provides the essential details of the system for both business and technical stakeholders. Expectations are expressed and managed using functional requirements" (par. 5).

"Who will use the product?" or "How will it be used?" are a couple of key questions. Andy Gurd (2013) suggests the following outline for functional requirements:

1. Introduction (tell readers what the document is about, i.e., the product, users)
2. Purpose (explain what the document does, i.e., functional requirements for the product)
3. Terminology (define anything that might not be understood by all readers)
4. App description (describe product)
5. System components and overview (describe key parts of the product and how they function separately and together)
6. Functional requirements (highlight what system must have or do to be functional for users)
7. Graphical user interface (GUI) (screenshots, drawings, etc., of prototypes of the product to show how it would function)

BOX 5.1 FOCUS CONCUSSION DETECTION SYSTEM REQUIREMENTS (COURTESY OF EYEGUIDE)

FOCUS CONCUSSION DETECTION SYSTEM 1.0 REQUIREMENTS

Overall System Activity

- Database cloud sync
- Authentication
- Pretest distance from screen locked to 381 mm using the scene camera
  - Need to figure out what this is and perform the proportion calculations to abstract out different screen sizes and resolutions
- Ensure no head movement during test also using the scene camera
- Stimuli are crosses with a clear center
- Add time stamps to frames to reduce synchronization error
- Three tests: smooth pursuits, saccades, antisaccades

Interface (Form Fields)

Basic subject data entry as discussed previously: Name, birthdate, previous concussion history, etc.

Tests

Smooth pursuit test:
- Figure 8
  - Stimulus blanking for 1 second, at two intervals, 0° and 540°
  - 10 loops

Workflow:
1. "Pupil locked," "now tracking your eye movements."
2. "Please keep the screen about 15 inches away from your eyes," system gives feedback until proper distance is achieved.
3. "Please follow the target with your eyes as it moves around the screen."
4. "Task completed, thank you."

Saccades test:
- Cross appears 5° to the left of center
- Blank stimulus one frame before it moves
- Movement to 5° right of center
- Then vice versa: right to left
- 1 second between presentations
- 20 presentations

Workflow:
1. "Pupil locked, now tracking your eye movements."
2. **"Please keep the screen about 15 inches away from your eyes," system gives feedback until proper distance is achieved.**

3. "Please jump/**keep?** your eyes to the target as it flashes back and forth on the screen."
4. "Task completed, thank you."

Antisaccades test:
- Cross appears in center of screen, stays up during entire test
- Target flashes for 1 second 5° left, user meant to look 5° to the right
- Next presentation 2 seconds after they return to the center cross
- Randomize left versus right presentation
- 20 presentations

Workflow:
1. "Pupil locked, now tracking your eye movements."
2. "Please keep the screen about 15 inches away from your eyes," system gives feedback until proper distance is achieved.
3. "Keep your eyes on the image in the center of the screen. A target will flash to the left or right of the center image. Try to move your eyes the same distance, but to the opposite side of where you saw it appear. For instance, if you notice it flash on the left side, jump your eyes to the right. If you notice it flash on the right side, jump your eyes to the left. Then move your eyes back to the center image until next target flashes."
4. "Task completed, thank you."

1.0 Visualizations:
- Instant replay
- "Spider"
- Point by point horizon series

Quant presentations (descending order of priority; priorities 2 and 3 are not 1.0):
Priority #1: Smooth pursuit test ("figure 8")
- Gain/velocity delta (less = better)
- Saccades (fewer = better)
Priority #2: Prosaccade test ("look at the dot")
- Reaction time (lower = better)
- Saccades (fewer [>0] = better)
Priority #3: Antisaccade test ("look away from the dot")
- Reaction time (lower = better)
- Saccades (fewer [>0] = better)
Errors/saccades in the wrong direction (fewer = better)

As an example, Box 5.2 shows how the Focus concussion detection system functional requirements would look in Gurd's (2013) suggested format. For brevity's sake, all detail shown in Box 5.1 is not repeated here.

Whatever format you choose or detail you can provide, dependent on where you are in the process, you need to know what your product can or should do to function

BOX 5.2 GURD'S FORMAT FOR FUNCTIONAL REQUIREMENTS (COURTESY OF EYEGUIDE)

FOCUS CONCUSSION DETECTION SYSTEM 1.0 REQUIREMENTS

Introduction

These are the 1.0 functional requirements for EyeGuide Focus, a proposed concussion detection system. First date for these requirements is October 15, 2013. We have gathered feedback from competitor research, internal discussions, and interactions with potential users, that is, athletic trainers, through interviews and rapid prototyping of mockups.

Purpose

The Focus concussion detection system enables qualified health care professionals, including athletic trainers on the sidelines of sporting events, to administer fast, efficient, and reliable concussion tests to athletes in a wide range of sports in which the potential to experience a concussive brain injury is prevalent and the present means of assessing such injury is subjective. To be easy to administer and work in any environment and also to be reliable and fast, Focus must have the following components:

1. EyeGuide Mobile Tracker headwear for capturing eye movement.
2. iPad or similar portable device for easy setup to provide a visualization that players can view on the sidelines.
3. Cloud-based data collection and dissemination so player information can be stored and accessed as necessary, and then after testing, including concussion detection, the player's results can be viewed after the fact for ongoing rehabilitation.
   a. Cloud also allows for all data to be collected so that the system's quantitative accuracy improves per test.
4. Existing literature on concussion detection must be converted into visualizations and scoring algorithms—we're not concussion experts! We are converting the science into software for detection.
5. A stable, nondistracted environment is crucial—we don't know how to do that yet but Mobile Tracker headphone will cancel out noise.
   a. Maybe a hood?
   b. Test away from sidelines?
   c. Chin rest for stabilizing?
6. Software has to display visualizations and has to be easy to pull up athletes quickly, carry out tests quickly, and review results quickly.

Terminology

- Focus is the concussion detection system prototype
- User is athletic trainers, anyone using Focus
- Lazy 8 is the visual pattern athletes will follow on the screen to detect concussion

App Description

Focus is easy to administer, works in any environment, provides instant results, and makes use of more objective technologies and protocols to determine the status of the athlete. Here's an example of how it would work: A football player comes to the sidelines and complains of symptoms that subjectively may point to a concussive brain injury. To be sure, the athletic trainer brings the player over to the Focus system. The player puts on the headwear, which is adjustable to fit almost any head size. Ear phones are built in to cancel out any distracting noise. The player looks at a screen. This can be a laptop, an iPad, or other tablet; the Focus system works with any device. As the player watches the screen, a small camera built into the arm of the headwear tracks the user's eye. A built-in scene camera as well as an accelerometer prevent head movement from disrupting the accuracy of testing. The player is instructed to follow a series of points of light on the screen. These "pursuit" tests are very brief and represent an automated version of tests given in ophthalmology clinics. They have been validated by prior research to be reliable determiners of concussive brain injury.

System Components and Functional Requirements

- Database cloud sync
- Authentication
- Pretest distance from screen locked to 381 mm using the scene camera
  - Need to figure out what this is and perform the proportion calculations to abstract out different screen sizes and resolutions
- Ensure no head movement during test also using the scene camera
- Stimuli are crosses with a clear center
- Add time stamps to frames to reduce synchronization error
- Three tests: smooth pursuits, saccades, antisaccades
- Basic subject data entry as discussed previously: name, birthdate, previous concussion history, etc.

Screenshots

See attached

as something users can use. This also serves to limit expectations to something manageable. EyeGuide, for example, knew that, for a 1.0 version of Focus, it couldn't deliver priority two (prosaccade) and three (antisaccade) concussion tests because it didn't have the data necessary from users in a study population to create effective "spider" and "point-by-point horizon series" visualizations.

So for the readers of the functional requirements, it noted that priority two and three were not going to be 1.0 deliverables. However, to track that they were desired functional requirements at some point, they were included. Your functional requirements will change as you move further along in your design and get feedback from users. And even for these functional requirements, including users on your team, allowing them to react and provide feedback to what you are building for them through focus groups, and engaging in affinity diagramming sessions, among other methods, is an effective UCD approach.

5.2 ENVIRONMENTAL ANALYSIS

We cannot emphasize enough just how important it is to design a product that works for users where they will use it. This means you must analyze the use environment, and obviously you must go there to do that. You also need to involve users. The site visit method in Chapter 4 describes in detail how to do this. There are other ways as well. But users actively participating in the design process, interacting with rough prototypes, or even just participating in structured discussions on location where the product will ultimately be used will provide you critical feedback that could very well mean the difference between a successful or unsuccessful product.

We've already provided a few examples of the importance of this, but here's another one. When EyeGuide put its first product out on the market in 2011, it intended to make eye tracking as affordable as possible to consumers. Previous eye tracking to that point had been prohibitively expensive for most researchers, costing between $40,000 and $60,000 to purchase. EyeGuide's first product was under $2000. But to be affordable, it cut corners. For one, it was wearable, allowing a less expensive camera to be used. It also ran on three AAA batteries. But these batteries had to be rechargeable because rechargeable batteries kept their power at the required operating level longer than standard, nonrechargeable batteries.

But in the real world, most people will find batteries lying around to meet their needs because (1) most battery-operated appliances run on nonrechargeable batteries, (2) most people just buy batteries that are nonrechargeable, and (3) most appliances are fairly lenient on energy requirements. People will take somewhat used batteries from one device they aren't using to power one they are using. More than once we've seen parents, in desperate situations because a child's favorite toy has stopped working, furiously break down and steal batteries from a remote control and then jam them quickly into the toy to bring it back to life.

But the EyeGuide 1.0 tracker was very specific. It required three fully charged, rechargeable batteries. As an early result, dozens of customers bombarded EyeGuide customer support with questions and complaints primarily related to battery issues. EyeGuide could have made a product that took nonrechargeable batteries. Its early prototype, in fact, did run on nonrechargeables, but at some point along the way, engineering decisions (technology-centered) took precedence over ease of use (user-centered). EyeGuide eventually created documentation to ship with future 1.0 trackers, and when that still didn't fix the issue, it began to include with each 1.0 tracker shipment a package of rechargeable batteries along with a charger—at no extra cost to the user. In other words, instead of finding out how users would use the product before making it, EyeGuide learned after the fact, and that decision not only caused it headaches and frustrated users, but also cut into company profit.

The level of environmental analysis will depend on the product and how it will be used. Go to where users work and live, and while there, ask key questions about the environment that you can then consider when designing the product. To make it simple, consider the senses, or at least four of the five (we can probably leave out taste), and ask the following questions:

- Is it loud where the product will be used? What kinds of noises will be around users? Which noises will be distracting? Which ones do they have to hear as they use the product?
- Are users engaging with others as they use the product? Will these conversations distract them? Will they use the product to facilitate these communications? How many people at once, and how often, will they talk?
- Will users be putting their hands on other things or carrying out tasks with other products while they also use your product? Will they hold your product? Will they wear gloves or have other physical limitations that impact engagement with your product? Will they be moving around while using your product? If so, where, and does the type of environment change as they move around? If they are relatively stationary or in the same environment, are they sitting or standing?
- Do users work alone without physical contact with others, or do they see people, even engage with them, while they work with your product? Are they in a cubicle or in an office? Are they looking at other things while they look at your product?
- Is the environment dangerous? Must the user do fast things, or stressful things, with your product? If it is an onscreen product, such as an application, what things about the users' screens (such as size, applications open, etc.) may impact your product's use? If users have offices or workspaces, what is in that space that distracts them but might also be used to carry out tasks?

There are many other questions to ask. If you can, take photographs of your users' environmental spaces so that you can actively consider it when designing. You can't accommodate every environment, but if your primary user audience uses your product most often in one kind of environment, your product's design must take that into account.

Another approach to understanding the use environment, and keeping that understanding active by updating it throughout the design process, is to create working models of the user environment, all fueled by direct user engagement, that represent different thematic aspects of how users practice what they do and how the product should, according to Holtzblatt and Beyer (2013), "support and extend its users' work practice" (sec. 1, par. 2).

The idea behind the five different models Holtzblatt and Beyer (2013) advocate is that they offer differing perspectives that, when taken together, capture effectively a well-rounded view of the user experience in the situation of use:

1. Flow model (shows communication between different people in an environment to get work done)
2. Cultural model (shows the culture of the environment that influences how people work—also notes points of conflict, tension, or barriers that keep users from doing their work)
3. Sequence model (shows how people do their work, the steps involved, their goals, and problems with getting work done)
4. Physical model (shows literally the physical environment of where people work as well as how people organize that environment to do their work)
5. Artifact model (shows what's made because of the work people do as well as what people create to help them work)

Holtzblatt and Beyer (2013), as part of a contextual design approach they have championed since the 1980s, provide helpful diagrams of each of these models. We encourage you, as you think about the user's environment, to create your own. To aid with that, we have included examples of these models (Figures 5.1 through 5.5) that EyeGuide used when beginning early work on analyzing the Focus use environment.

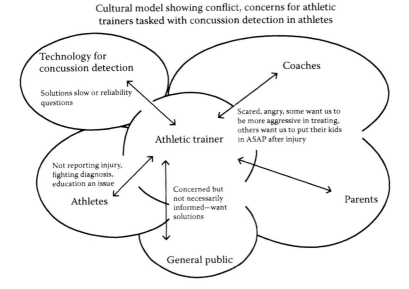

FIGURE 5.1   Cultural model of Focus concussion detection system. (Courtesy of EyeGuide.)

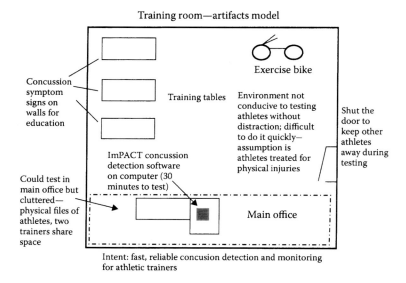

FIGURE 5.2   Artifacts model of Focus concussion detection system. (Courtesy of EyeGuide.)

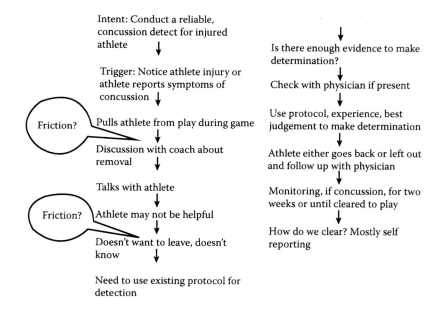

FIGURE 5.3   Sequence model of Focus concussion detection system. (Courtesy of EyeGuide.)

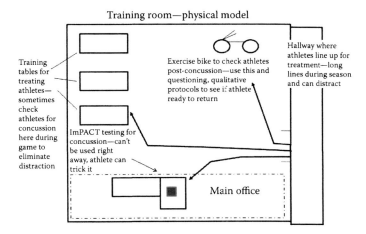

FIGURE 5.4   Physical model of Focus concussion detection system. (Courtesy of EyeGuide.)

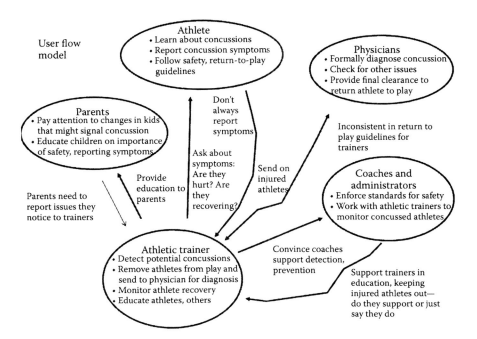

FIGURE 5.5   User flow model for Focus concussion detection system. (Courtesy of EyeGuide.)

All of these models allow you to visualize the situation and then discuss challenges for the product to be effective in that situation with as many stakeholders as possible, including users. It's what Holtzblatt and Beyer (2013) call "user environment design" (sec. 1, par. 4). What you are essentially doing is rendering to an operable form the process of making something work in the real world. It's one thing to say you are going to do UCD to make a product work; it's another to devise techniques and create deliverables that allow you to demonstrate your increasing awareness of the users and the situation of use. We create models like this, or we generate other documents, such as the functional requirements, so that we know we are creating a real product to work in a real space. The drawings, the prototypes, the checklists, and other documents catalogue our work and also serve as a means to get feedback from users. We can talk with users about the culture of the workplace, for example, as we show them a diagram of the cultural model as well as a prototype of a product we think might make the culture better. EyeGuide frequently talked with athletic trainers when beginning to develop Focus. It learned that athletic trainers faced many conflicting pressures ranging from athletes who underreported their symptoms to coaches who wanted the kids to be returned to play as soon as possible. EyeGuide's modeling of the user environment allowed it to understand that a product was needed that worked to deliver not only fast results, but also reliable results that coaches, players, and even parents would accept.

An effective UCD product should make the users better at what they do. It should improve communication, eliminate or offer solutions for cultural barriers, and it should speed up work and make it better. Users shouldn't have to make the product fit the environment. The product should fit and also improve the use environment. How can you design a product to do this? It has to be functional, but it also has to be integrated into the situation. Evaluate that situation carefully, create a depiction of it to serve as an ongoing reminder, and, whenever possible, have users use prototypes of the product in that situation so their interactions will help inform changes to the product that make it a better tool for users where they work and live.

5.3 ORGANIZATIONAL ANALYSIS

As the UCD designer, your job is to operationalize your organization's goals. That is, what an organization wants is not necessarily the same thing as what you're tasked with. For some organizations, such as startups that make mobile applications to be sold in the iTunes store, goals and designs are basically one in the same. The organization, in other words, is the application. But in other organizations, you have to work a little harder at making your product's design fit not just user wants and needs, but also organizational aims. If you're designing a new website for a nonprofit organization, this means making sure you understand what the organization wants the website to say about itself in relation to its users. Your job, once you figure that out, is to create features and include content that enable the organization to get across its message while, at the same time, meeting the wants and needs of the users. Not so easy to do, huh?

Adding to the problem is that many of an organization's goals are not necessarily focused. Or an organization will have way too many goals. You need to work with stakeholders and clients to figure out which goals matter most and then focus on building something that realizes the most important goals and also achieves user goals.

Pratt and Nunes (2012) suggest creating a priority pyramid in which stakeholders use a modified affinity diagram exercise (see Chapter 4) to put up sticky notes of goals on the pyramid with high at the top, medium in the middle, low on the bottom. This will promote discussion and practicality. Eventually, only so many goals can be on top with the highest priority. Take a look at Figure 5.6 and how EyeGuide used Pratt and Nunes' advice to craft a goals pyramid and devote its efforts for the 1.0 version to fulfill just the high-priority goals.

A pyramid, as you know, is smaller above than below; smaller means more focused, which usually means a better deliverable. Think again of Garrett's (2011) idea of the multiple planes of interaction. By figuring out organizational goals early on, the scope part of the equation can be solved. More than once, a product has failed because users didn't like features that were created because the scope was too big and demanded the product try to do too much. Even if you are designing on your own, you have to make sure your goals don't overwhelm product scope.

Along those lines, scope creep, or when the focus of your project's design changes, can be detrimental as well. When goals change, requirements change, and deadlines, budgets, and product quality can all be impacted severely. Years ago, Brian was a web developer tasked with building an online transaction system for a government entity in Minnesota. This was in the relatively early days of the commercial web, around 2000, and at that point, there weren't a lot of mature, reliable third-party vendors offering online transaction services. Brian was tasked with building one of his own, intended to provide a means for cities and other local government agencies in Minnesota to make online payments to Brian's organization for a variety of services, such as legal and insurance fees.

The project was supported at the highest levels, and Brian was given money to hire an outside developer to work with him to create code to take transactions, process them, and then reconcile them so that the exchange of money from customers to

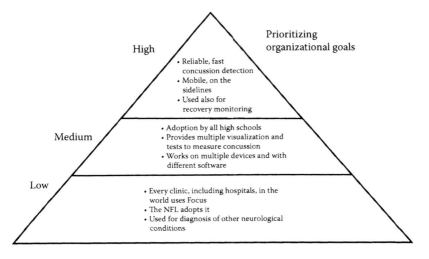

FIGURE 5.6 High, medium, and low goals prioritization. (Courtesy of EyeGuide.)

the organization would be as secure and automatic as possible. Brian met with stake-holders throughout, worked through the steps for how the transactions would work, and also consulted with reluctant managers to reassure them that this new transaction system would ultimately mean less work, not more. Brian even designed the system to handle automated clearing house (ACH) transactions, not credit cards, because the finance director and his team were more accustomed to and better trusted ACH, more than other purchasing tools.

As the project developed, changes began to creep in. First, the original develop-ment language, Cold Fusion, was abandoned in favor of Java because the organi-zation was moving all of its development to that platform. Brian, as a result, had to bring in a different developer with the right skills, which cost time and money because a good percentage of development had already occurred in the previous language. Finally, the project was nearly completed when, during last-minute testing, the finance director's staff reacted negatively to having to use a new interface rather than their present one (which was not forward or customer facing) to view and recon-cile transactions. Effectively, the project died right there because it was out of money and a stakeholder had introduced a new requirement. Therefore, despite the best of intentions, scope creep not only (initially) stalled the project, it eventually killed it. To this day, the organization in question still doesn't use an internal or third-party vendor for transactions. Customers continue to pay, literally, the old fashioned way, by checks through the mail.

Get the stakeholders to agree that although there can be change based on user feedback and ongoing product development, if the overarching business goals change, the project's scope is in jeopardy. Save medium and low goals for the next effort or a 2.0 version of the product. EyeGuide made this choice. For what it needed to accomplish, multiple visualization tests were not necessary. Nor was it neces-sary that Focus worked on multiple devices. Focus could be a successful product for EyeGuide and user-centered at the same time just by focusing on high-priority goals during its 1.0 development.

Goals are not the only things to consider when analyzing an organization. What about money? How much do you have to get the job done? If you are doing this as a contractor, you need to consider how much it will cost you to get the work done. As part of assessing the situation, what will it cost to build the product, and will the organization give you this amount to do it? If not, you need to consider, at the begin-ning, ways to change the product's scope so that it matches what you can afford to make. There are many resources, such as Russ Unger and Carolyn Chandler's book, *A Project Guide to UX Design: For User Experience Designers in the Field or in the Making* (2009), which offers more in-depth guidance on how to manage the financial details of a design project. But everything costs something. Your time, should it be donated for free, is still an expense because it means you're not using that time to make money. We often think about this in terms of hours, but one of the reasons why we employ RABBIT is to make sure we are aware of everything that needs to be done to deliver an effective UCD product. Time management, along with money, is part of that. What resources will your organization provide you to do the work? How much time? How much money? You need to know this as you start the process, and you need to monitor it as you move through to conclusion.

5.4 COMPETITOR ANALYSIS

At some point early on, in addition to determining user goals, client goals, and the overall requirements of the product, you'll also have to understand what competitor products are out there in the same space. This is invaluable information because it enables you to learn from others what does and does not work. Once the analysis is complete, you'll also have a better sense of what your product needs to do to occupy what is sometimes called white space or an area in the competitive landscape not presently owned by competitor products.

First, identify and list any products on the market now that are more or less direct as well as possibly indirect competition. If you are doing this for your organization, get help from your client or marketing folks internally and also do some research of your own via the web, even visiting stores, watching users in action, etc. What is the product? What is its cost? What does and doesn't it do? What do people say about it? Does it do anything unique?

From there, consider using a graph (Figure 5.7) for displaying all of the competitors in a space or landscape. For example, when looking at competitors in the concussion detection space, EyeGuide created a grid with a y-axis (vertical) for cost, ranging from high to low, and an x-axis (horizontal) for functionality, ranging from great to bad. Obviously, great, acceptable, and bad are subjective labels and open to interpretation, but that's okay. The point of this exercise is to find out where your existing product is in comparison to competitors or where you'd like to position your new product so it will be competitive against other products already on the market.

To do this right, you put all of your competitors into spots where you think they best fit given the ranges you've identified. Don't make your competitors out to be worse than they are. If anything, give them more credit than they possibly deserve. You don't want to cheat here because the reason why you're doing this exercise

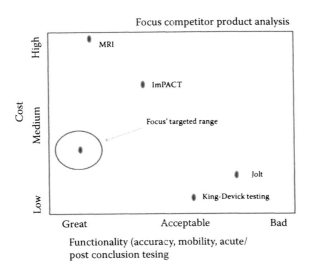

FIGURE 5.7 Focus competitor grid. (Courtesy of EyeGuide.)

is to determine how best to place your product so that it can compete effectively against other similar products. You might circle or make larger those you think are more successful than others. What you're attempting to do is conceptualize (1) what characterizes, from a big-picture perspective, a successful competitor product; (2) how similar competitor products are (if a lot are clustered together that tells you something); and (3) what space exists for you to enter the market and be successful.

Once you've done this, you can begin to determine what you need to have in your product that not only achieves business and user goals but also, once it goes to market, has a chance to be competitively successful because it does things, or occupies a space not currently occupied. This graph enabled EyeGuide to understand (given its technology costs) that it would not be able to compete on an equal footing, in terms of pricing, with a low-technology product like the King–Devick Test. But, as a goal, it could have a place as near to it as possible so that users of King–Devick or others looking for an affordable price would be interested. At the same time, the graph showed EyeGuide there was a strong position to be competitive in the market if it could be somewhat affordable and return, at the same time, great product functionality.

Obviously, you don't want to be so unique that you're literally off the competitive map. This sort of forward-thinking, cutting-edge product has a hard time catching on. There's something to be said, from a business and user-centered design perspective, for making a product that capitalizes on what your competitor products do right while trying to do a few key things that are novel or better in design and implementation.

A great technique for a final visualization of your product's white space is a Venn diagram (Mukund Mohan's blog, *Best Engaging Communities*, 2015, demonstrates this and other useful methods for displaying your competitive landscape). Each circle in the diagram represents a key service or feature. Put competitors into the circles of the Venn, appropriately overlapping the circles because an ideal product would deliver something that allowed it to exist in all the circles. If you look at Figure 5.8, you see that EyeGuide created three overlapping circles of cost, mobility, and accuracy. What it noted, as Figure 5.8 demonstrates, was that there was white space for a product that could deliver on all three because almost all other competitor products delivered on and only occupied one or two of the spaces or circles.

The Apple PC Lisa is a great example of what happens when you build something that is so far beyond what users understand or can use that it fails even though, in many respects, it was a well-designed, highly functional, if not groundbreaking, product. Jobs and Apple debuted Lisa in 1983 to much fanfare. It was created to be revolutionary, and in many respects it was. The first graphical user interface (GUI) computer for people to use at home or at work, it offered some never-before-seen features, including how it handled documents. But it had its quirks. It wasn't built with a true understanding of what users needed (Lumb 2015), and it was really expensive—almost $25,000 in today's dollars. Jobs basically lost his job over it (he did get it back), and Apple ended up only selling around 10,000 units.

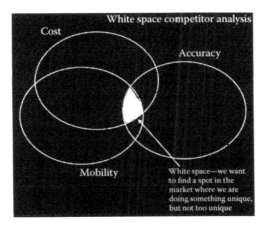

FIGURE 5.8   Focus white space Venn diagram. (Courtesy of EyeGuide.)

You really can't blow users minds too much. In other words, there is such a thing as too much white space. Yes, you run the risk of making just the same thing and failing that way, and if you go on just what people say they want well, then, you'll just end up making a faster horse, not a car (see Chapter 1). Ultimately, it is a delicate balance to do UCD effectively, and analyzing the competitive landscape helps you with this. Whereas you can't introduce a product that is out of reach, financially or conceptually, for its intended users, you also, to be as successful as possible, need to create a product that does what it does a little differently and a little better than what competitors do. Your competitive analysis helps you identify what people like and understand as well as, conversely, what they don't like and understand. This can be defined not just by cost or interactive features, but also include color, size, structure, and content. The goal is to build something that takes advantage of what people can do or know how to do and then push them to do more things or be better at what they do in ways that are understandable and thus effective. Again, we are attempting to bridge that gulf Norman (1986) writes of. If we don't make something a little different, we run the risk of it failing because no one will adopt it; on the other hand, if we make it too different, it may fail because people don't know how to or cannot adopt it. A competitive analysis helps us negotiate this challenge.

EyeGuide learned through its competitive analysis that its approach to concussion detection, measuring the ability of a person to track an object visually, was borderline almost too different for its users. The science indicated that this approach was reliable, if not more reliable, than other approaches. But the process of using an eye tracker, explaining how it worked to detect concussion, was a barrier to entry. EyeGuide couldn't abandon this approach, however, because it was the core technology of Focus. What it could do was eliminate most of the confusing or intimidating aspects of eye tracking, creating a headset platform that was as easy and fast to use as possible. Also, by putting most of the interaction with the product in the iPad software interface, EyeGuide took advantage of users' existing familiarity and comfort level with iPads and touchscreen manipulation. Users understood and liked to use iPad interfaces, and that softened or made more approachable EyeGuide's more disjunctive or confusing aspects. Competitive analysis, in other words, helped EyeGuide understand what it needed to be like and not like competitors to make an effective UCD product.

5.5 MATERIALS ANALYSIS

Let's assume you know your users, you know their goals and your own goals, and maybe you even know where the users will most likely use your product and what about that environment will influence that use. But you also need to spend some time understanding what about the product's materials will influence how it will be used and what can be done within the constraints of those materials to deliver a usable product to users.

We've previously used the example of an online physician reference tool to explain the value of understanding where users will actually use the product and how that environment matters. But part of the issue with the way the physician reference tool was built was related to an ineffective use of materials by the developers. They forced the users to interact with content and features on a mobile device that was, in fact, built for use on a laptop or desktop, something that was typing-heavy and that expected large screens for viewing. A mobile device has a gestural interface and a small screen, at least compared to a desktop monitor, and although users can type, it is done much more slowly. If you remember, one physician during testing of the reference tool actually blamed himself for inefficient typing, saying he suffered from "fat fingers disease." But he wasn't the problem. The problem was that the materials of the product weren't considered when the product was designed.

What kind of device will you design for? At minimum, you must match features to materials, user to environment. You should consider a range of factors, comparing them to user needs and the space where they will use the product:

- Keyboard size, type, number of inputs
- Screen size
- Product weight, overall size
- Product stability
- Portability
- Storage capacity
- Video, audio
- Security
- Touchscreen
- Voice-to-text options

Spend time understanding how the product will be used. Again, the best way to do this is to create prototypes of it early to get real user feedback. You can also watch users in action using similar products. How do they use it? What do they like? What do they do to make things work? Can you tell what causes them problems? How can you design something that makes them successful?

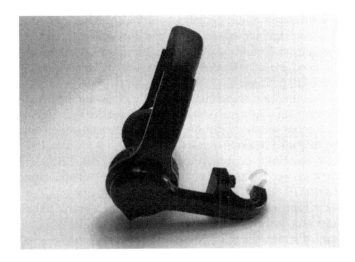

FIGURE 5.9   EyeGuide mobile tracker. (Courtesy of Dr. Brian Still.)

EyeGuide knew that its headset would cause issues for some users. Although it is not the long-term solution for future concussion detection systems that will be put out by the company, the headset was the way EyeGuide collected visual data from wearers in the earliest stages of Focus' development, and it often proved a challenging fit. Some football players had head sizes that stretched the headset dimensions, which originally were created for wearers with head sizes ranging from US 4–8. The hot mirror (Figure 5.9), which was used to collect a reflection of the user's eye from the camera that tracked and gathered visualization data, was also somewhat tricky to put into the right position for accuracy, especially when users wore glasses.

One accommodation, however, based on a material analysis of use in the actual situation, was to purchase a chin rest. This stabilized the users' heads and made for better data collection. Also, the eye camera's software used for collecting the data was reconfigured to capture a clearer image of the eye, speeding up and also making more efficient the process of setting up users for concussion testing. It is often the case that you cannot mitigate every issue regarding the product's materials, but you can make accommodations. Conduct an exhaustive analysis of the materials of your product, what can and can't be done or done well, and that analysis will help you in crafting a more usable product.

5.6   CONTENT ANALYSIS

Although you can choose to do any of the previous analyses in basically any order you prefer, we encourage you to perform this last one, content analysis, at the very end of your assessment of the situation. Your sense of the content, images, words, graphics, etc., will be better informed by what you learn conducting your other analyses; it will also serve as the perfect bridge method to the next part of the RABBIT process when you balance all the competing concerns, such as your user needs and wants as well as environment, material, and functional factors, and then filter what you want the product to do, and even what the user may want the product to do, with what can be done and needs to be done, realistically, given all considerations, to make an effective UCD product.

Typically, content analysis, sometimes called a content audit, occurs when you evaluate the content of an existing product, such as a website. You will try to determine how much content exists on the site and then also judge how much of that content is good or should be carried over to a new version of the site. But we believe content analysis, especially as part of your assessment of the overall situation surrounding the new design of a new product, makes sense even if no content, not even a product, yet exists.

Content is the means by which we communicate with users. It puts into action all of the messages, concepts, and functions we have decided are important for users to understand and then use to carry out whatever they want or need to accomplish with the product. We could guess what those are or wait until we begin the design of the product to consider, but that runs the risk of trying to use too much content or, when we test the product, evaluating the wrong content. By this point in the process, we know who our users are and where they will use the product, just as we know what competitors are doing that we need to emulate or counter. What we don't know is the content we'll need to satisfy user wants and needs to be competitive and user-friendly in the situation of use. A thorough content analysis will help us with this.

As Jones (2009) points out, "Content analysis results in a clear, tangible description of your content...Content analysis provides the foundation for comparing existing content with either user needs or competitor content, letting you identify potential gaps and opportunities" (par. 1). Detzi (2012) suggests such a content analysis or audit can be quantitative as well as qualitative, and doing both makes sense. How much content do I need to make an effective UCD product for my users? What kind of content? What's the balance in images and words? What about colors and interactive features?

The medium of the product, such as a mobile device, and even the culture or language of the user will impact your content decisions. In later chapters, we focus in depth on these concerns. For now, at this stage of development in the RABBIT process, you want to make your first attempt at listing the content you think needs to be included in your product. The detail of what you create can vary, and yes, will change, as it should, based upon the work you do going forward. Draw on other analyses you've done as well as your previous user research to create an artifact that immediately can be shared and discussed, even revised, by members of your team and also, most importantly, by your users who are helping you design. Involve users, again asking them questions about your proposed content, allowing them to participate in affinity diagramming or other group work exercises.

What you end up with may just be a simple listing of content descriptions without anything completed in detail, like the example in Figure 5.10.

Visuals	Language
Live video of eye for verifying calibration	Instructions for athletic trainer
3-2-1 countdown	Language for users? Not too much
Green checkmark for verifying good test	Report test score, provide written alternative to visual results
Red warning for bad test	Form field for entering user demographic data
Moving reticle for visual tracking test	Reports of test result, listing of symptoms to be selected
Replay results of test	
Visuals showing quality of test, good and bad	

FIGURE 5.10   Simple content analysis for Focus concussion detection system. (Courtesy of EyeGuide.)

Or you may opt for something that is more detailed. Leise (2007) discusses a number of content inventory heuristics to ensure you have included content that meets the standards of what is considered user-friendly:

- Collocation (collect relevant content in same place—should be easy to find)
- Differentiation (don't put different content in one area; put in own areas and label as different)
- Completeness (have just enough content but no more)
- Information scent (create logical labels)
- Bounded horizons (make sure users know where they are and what it means)
- Accessibility (make sure users can access the content)
- Multiple access paths (give users different ways to access, view, and use content)
- Appropriate structure (match content and organization to what users expect, their mental models)
- Consistency (don't change how you deliver content or users will get confused)
- Audience-relevance (make sure everyone, even with different needs and backgrounds, can use the content)
- Currency (use current content)

Using Leise's (2007) heuristics, Figure 5.11 shows a more detailed content analysis that EyeGuide used to determine the content it needed in Focus before moving forward with designing a prototype. You'll notice it is rough and not definitive, but it serves to crystallize team thoughts at that point, given the research done, for what content would be required to make Focus effective. Should you want, you could forego writing this out in tabular format and instead begin to sketch out mockups or put the content into some sort of structure form. You'll still need to filter it based on need and other factors, but there are no hard and fast rules. Just carry out the analysis.

Focus content analysis by Heuristic	
Content Heuristic	**Solution**
Collocation (collect relevant content in same place—should be easy to find)	Organize user profiles by team, sport, also allow for searching
Differentiation (don't put different content in one area; put in own areas and label as different)	Need to explore this further??? But one idea is not to have different content. Let's focus on one thing so there won't be any need for differential content. Only can think of help now and we will make that in-context help so they can get it whenever they need it at the place where they are stuck
Completeness (have just enough content but no more)	Remember, we're on the sidelines or in a locker room. They have to do this quickly and repeatedly without error. Let's forget about major reporting, looking at large data sets from population. They only need the means to test, monitor one athlete at a time for 1.0
Information scent (create logical labels)	User profile with name, photo to match, have breadcrumbs consistent throughout for profile, home, results—need to have consistent terminology
Bounded horizons (make sure users know where they are and what it means)	This goes to labeling but trainers need to know when they are running a baseline test and when they are viewing concussion test results, how to get back to and view results of multiple tests
Accessibility (make sure users can access the content)	They will need to login but make this intuitive, and also make it easy for them on sidelines to find athletes quickly—search is big here, and also layout
Multiple access paths (give users different ways to access, view, and use content)	Search by player, by team, by date, by concussion result, and also give trainers data, hard numbers along with visuals (thumbs up, thumbs down, graphs showing scoring by color gradient, green=good, red=bad)
Appropriate structure (match content and organization to what users expect, their mental models)	Make this feel like an easy to use iPad, also use language for description, diagnosis that makes sense; remember too that eye tracking is hard so let's not make this eye tracking, let's make this an easy to use iPad app with all the same features and let's take pain out of the eye tracking part
Consistency (don't change how you deliver content or users will get confused)	1.0 is 1.0. We are sticking with this as a team organized deliverable, individual athlete selection, testing and monitoring. No change for 1.0
Audience-relevance (make sure everyone, even with different needs and backgrounds, can use the content)	We're targeting trainers now, but some are tech savvy some are not, and they're used to ImPACT which is text heavy. We need to balance text description with "touchable" iPad features for control
Currency (use current content)	Sync button big—let's encourage them to sync and make it easy so all scoring is up to date

FIGURE 5.11   Focus content analysis (Leise's Heuristics). (Courtesy of EyeGuide.)

TABLE 5.1
Assess the Situation Checklist

Check When Completed	Analysis Questions	Analysis Answers
☐	**Functional Analysis** What must the product do?	
☐	**Environmental Analysis** Where will the product be used, and how should we account for that so that the product integrates effectively into the environment if not actually enhances the environment for the product's users?	
☐	**Organizational Analysis** What are your organization's goals for the product? What resources will you be provided? What's the budget? What limitations or other factors will impact how you design the product?	
☐	**Competitor Analysis** What other products are like yours? What about yours will be different? What white space can you occupy?	
☐	**Materials Analysis** What will go into the construction of your product? How will this impact its design and how users use it? What products may users have to interact with what you've built for them that you need to consider?	
☐	**Content Analysis** What content has to go into your product to satisfy users' needs and wants? What can be left out?	

With the conclusion of your content analysis, you will have completed your assessment of the situation. Refer to the checklist in Table 5.1 as a final tool for verifying you have all you need to go forward in the design process.

TAKEAWAYS

1. Functional, environmental, organizational, competitor, materials, and content analyses will help you assess both the situation that surrounds your design process and the situation in which users will use your designed and delivered product.
2. Functional requirements should allow users to successfully accomplish real tasks. Features that don't accomplish anything for users should be downgraded in priority or removed.

3. Functional requirements will change as you move further along in your design and as you get feedback from users.

4. The level of environmental analysis that you need will depend on the product and how it will be used. Go to where the users work and live, and while there, ask key questions about the environment that you can then consider when designing the product.

5. Create working models (flow, cultural, sequence, physical, and artifact) of the user environment to help you understand how users practice what they do and how the product might support user activities.

6. An effective UCD product should make users better at what they do. It should improve communication, eliminate or offer solutions for cultural barriers, speed up work, and make work better. The product should fit and improve the use environment.

7. Your job as a UCD designer is to operationalize your organization's goals. You need to work with stakeholders and clients to figure out which goals matter most and then focus on building something that realizes the most important goals and also achieves user goals. Figuring out goals early on helps refine scope and avoid scope creep.

8. A competitor analysis helps determine where your existing product is compared to competitors, or where you'd like to position your new product so it will be competitive against other products already on the market.

9. You must match features to materials and user to environment. Compare user needs and the space where they will use the product.

DISCUSSION QUESTIONS

1. Compare the cultural model (Figure 5.1), artifacts model (Figure 5.2), sequence model (Figure 5.3), physical model (Figure 5.4), and user flow model (Figure 5.5) for the Focus concussion detection system. What does each of these models tell you about users and their needs? What type of information is most helpful from each model? What did you learn from one model that wasn't indicated in another model?

2. What can you do to prevent introducing a product that is out of reach, financially or conceptually, for its intended users? What method can help you create a product that does what it does a little differently, and a little better, than what competitors do?

3. What types of things should you consider when assessing material features?

4. If we cannot mitigate every issue regarding a product or its use, how can we make accommodations for this?

5. What might a completed assess the situation checklist look like? How would you communicate these findings to a client or programmer?

REFERENCES

Detzi, C. "From Content Audit to Design Insight: How a Content Audit Facilitates Decision-Making and Influences Design Strategy." *UX Magazine*. March 20, 2012, accessed February 20, 2016. https://uxmag.com/articles/from-content-audit-to-design-insight.

Garrett, J. *The Elements of User Experience: User-Centered Design for the Web and Beyond.* Berkeley: New Riders Publishing, 2011.

Gurd, A. "Managing Your Requirements 101—A Refresher. Part 4: What Is Traceability?" *IBM Requirements Management Blog*. January 28, 2013, accessed February 20, 2016. https://www.ibm.com/developerworks/community/blogs/requirementsmanagement/entry/managing_your_requirements_101_a_refresher_part_4_what_is_traceability7?lang=en.

Holtzblatt, K., and H. Beyer. "Contextual Design." In *The Encyclopedia of Human-Computer Interaction,* 2nd ed., edited by M. Soegaard and R. Dam, sec. 8. The Interaction Design Foundation. 2013, accessed February 18, 2016. https://www.interaction-design.org/literature/book/the-encyclopedia-of-human-computer-interaction-2nd-ed/contextual-design.

Jones, C. "Content Analysis: A Practical Approach." *UX Matters*. August 3, 2009, accessed February 20, 2016. http://www.uxmatters.com/mt/archives/2009/08/content-analysis-a-practical-approach.php#sthash.pWtIhyD9.dpuf.

Leise, F. "Content Analysis Heuristics." *Boxes and Arrows*. March 12, 2007, accessed February 20, 2016. http://boxesandarrows.com/content-analysis-heuristics/.

Lumb, D. "Remember Apple's Lisa, The Computer That Cost Steve Jobs a Gig? Kevin Costner Does!" *Fast Company*. January 22, 2015, accessed February 20, 2016. http://www.fastcompany.com/3041272/the-recommender/remember-apples-lisa-the-computer-that-cost-steve-jobs-a-gig-kevin-costner-d.

Mohan, M. "The Ultimate List of Competitive Analysis Landscape Charts with 7 Complete Examples." *Best Engaging Communities*. June 2, 2015, accessed February 20, 2016. http://bestengagingcommunities.com/2015/06/02/the-ultimate-list-of-competitive-analysis-landscape-charts-with-7-complete-examples/.

Norman, D. "Cognitive Engineering." In *User-Centered System Design: New Perspectives on Human-Computer Interaction*, edited by D. Norman and S. Draper, 31–61. Hillsdale: Lawrence Erlbaum Associates, 1986.

Pratt, A., and J. Nunes. *Interactive Design: An Introduction to the Theory and Application of User-Centered Design.* Beverly: Rockport, 2012.

Sankar, V. "Managing Your Requirements 101—A Refresher Part 2: How to Write Good Requirements and Types of Requirements." *IBM Requirements Management Blog*. January 14, 2013, accessed February 19, 2016. https://www.ibm.com/developerworks/community/blogs/requirementsmanagement/entry/good-requirements-writing?lang=en.

Unger, R., and C. Chandler. *A Project Guide to UX Design: For User Experience Designers in the Field or in the Making.* Berkeley: New Riders Publishing, 2009.

6 Balance and Filter Design Features

If you attempted to create a product that met every user need or want, you might end up with a 60-button TV remote control like the one Brian uses at home (Figure 6.1).

More options, more than Brian knows how to use, usually means that he ignores a lot of what is available to him or he makes mistakes, pushing inadvertently one button when he meant to push another. Barry Schwartz (2004) calls this the paradox of choice. If you bombard users with too many options to satisfy their or your goals, you'll overwhelm users and paralyze them into inaction. They don't see trees; they see a forest, and they don't know how to make their way through it.

Just because you have the money to make a product and you can do it and the user wants it, doesn't mean you should do it. On the other hand, you may be faced with an opposite situation in which you have very limited time and money to meet user needs. In other situations, there could be conflicts between what users need and what your organization wants to do or can do. Balancing needs is about prioritizing and filtering. You have to make hard calls most of the time, striking the right balance by giving priority to what will make the product work best for the user and, at the same time, achieving your organization or design goals.

Interestingly, this is something users already do themselves. Nielsen (2006) and others have noted that when viewing a website, users will scan what they want to see rather than follow the text or other features as the website has been designed for them to be read. Through eye-tracking technology, which allows researchers to see where users' eyes actually look, we see a pattern for scanning, in fact, that looks much like an F across the screen (Figure 6.2).

Squint your eyes a little bit, and you'll make out horizontal lines flowing across the page, extending away from a more brightly colored, yellow vertical line. Each splotch represents a user eye fixation or a brief moment when he or she looked closely at something on the page. More splotches, and correspondingly brighter colors, mean more intensity of focus. This is what is typically called a heat map. The more focus or the hotter the intensity of focus, the color of the heat map changes, going from gray or not much intensity to bright blue, then yellow, and finally even red for the greatest intensity of focus. Think of it like a Doppler satellite radar that shows inclement weather. Heavy thunderstorms, for example, are dark red in intensity compared to light rain, depicted in green. According to the site's design in Figure 6.2, the bright splotches should appear on the content, the users' eyes moving with the content as it flows from left to right.

But users don't read online as much as they scan. Moving quickly, they pick out what they want to see, and they don't look at what they think is not essential for them to use the page's content. On the website in Figure 6.2, like many others with a lot of content, the users scan horizontally across the top first to make the top part of the F.

FIGURE 6.1   60-button TV remote control. (Courtesy of Dr. Brian Still.)

They then scan again, horizontally, below the top line, creating two to three more lines of brighter activity. Finally, the vertical line on the left shows that they have spent time trying to orient themselves to the site's content because many website menus appear on the left. This F pattern made doesn't match up with the layout; instead, it represents the visual activity of users as they scan and read quickly for meaning.

The F pattern is repeated so often by users that some designers (Jones 2012) create interfaces that mimic it. Rather than fight the tendency of users to scan more than read online, such designers create layouts that put key content and other features in locations that correspond consistently with where users' eyes are fixating as they view and filter.

Such natural filtering on the part of users is not only reserved for how they view screen content. Even though the average user downloads 26 apps over time (Fox 2013)

FIGURE 6.2 Heat map showing F pattern of scanning website content. (Courtesy of Dr. Brian Still.)

to their mobile devices, a comScore (Lipsman 2015) report revealed that most users spend about half of their time using just one app; in fact, they spend almost 90% of their time using five or less. Users may download an app, they may say they want it, they may even pay for it, but ultimately they may not use it, or if they do use it, they may not use it as we anticipated.

What this all means for UCD is that our goal isn't to give users everything, overwhelming them with features. Rather, our goal is to give users what they want and need within their particular use situation to be effective and to work efficiently so that they derive some pleasure or satisfaction from the experience. Remember, users want to be in control, which is why, when they must, they will filter the apps on their mobile devices or the content on the screen.

6.1 FILTERING FEATURES

In the previous chapters detailing the RABBIT approach to UCD, you learned important information about your users, and you also learned about the situation(s) in which those users would use your product. You learned about the content users want, the organization's goals, the materials, and the time and money you have at your disposal to design the product. Now you need to take all of this information and filter it into what will be the set of requirements you use to build out a 1.0 version of your product. It will not be a product with everything that you can do; rather, it will be a product that includes everything it needs to do to be user-centered.

In *Willy Wonka and the Chocolate Factory*, a 1971 fantasy film starring Gene Wilder, the Everlasting Gobstopper is a candy that has changing flavors and colors and never gets smaller. It is the perfect candy, one with mysterious, multiple ingredients that are magically mixed together. If only the products we make for users could be like this. However, like the Gobstopper, that is pure fantasy. Hard work goes into balancing organizational needs with user needs, designer wants with user wants, and production capabilities with users' situational exigencies. You're not Willy Wonka.

But you do have a simple, yet effective tool at your disposal: triangulation. Just as you used a triangulated set of methods to understand your users, you should use triangulation again to filter down your full set of possible product features into a reduced set that meets user and organizational goals, all while working in the situation in which users will use it. You can't just adhere to organizational goals for your filtering. The result will be an organization- or technology-centered product. You also can't build something that matches every user wish. It isn't user-desired design; it is user-centered design. Finally, you can't build it just to match the situation and not consider user or organizational concerns. The result of that is a smoking cessation kiosk that no one wants or knows how to use.

Another way to conceptualize filtering is to think about how water is treated to go from what starts in a reservoir but ends up in a glass to drink. You could drink right out of the reservoir but run the risk of swallowing various impurities, including bacteria, not yet filtered. What you want to drink that satisfies your thirst, safely, has to be filtered. Water treatment facilities use a lot of techniques and processes to do this. Aluminum sulfate is added to cause the harmful particles to form together into sludge that's then caught and removed as the water is moved rapidly through sand and gravel. Chlorine and ammonia are finally added, and the water is tested one last time before it heads into pipes that send filtered, healthy drinking water to your house and eventually your glass.

UCD filtration really works in much the same way, minus the chemicals. Take a look at Figure 6.3 for a visual illustration. At this point in the RABBIT approach, you have a full set of features. You have a lot of information and a wish list of possibilities. But all of that has to be filtered. To do that, you must pass through the items on your wish list through user goals and needs, organization and even client goals and needs, and finally situational factors. If the feature you started with in your wish list moves its way through all of these filtering processes, it should remain in the final, reduced set of 1.0 product features. If it does not make it all the way through, strongly consider keeping it out. Don't get rid of it. Water treatment facilities often use the sludge for fertilizer. Eventually, you may be able to do something with the features that don't make it. However, to deliver a balanced product to users, only include in it those features that make it through the entire process of UCD filtration.

Brian and his lab were tasked with testing a new website for a university's library. The library development team had done a fairly thorough job of investigating what they thought were their primary user groups: faculty, graduate students, and undergraduate students. Most of their research to come up with these users as well as the features they indicated they wanted on a new library website were primarily preferential and gathered using self-reporting methods, such as surveys, focus groups, and

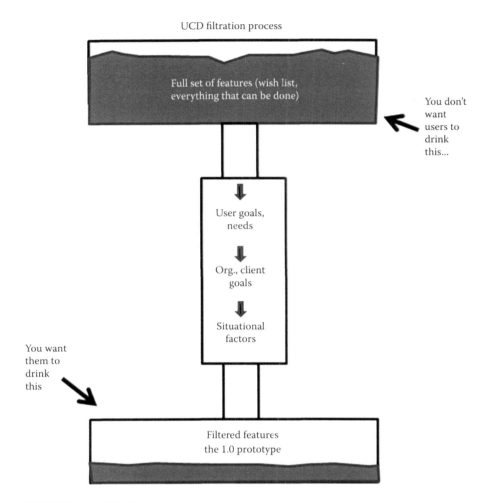

FIGURE 6.3   UCD filtration system.

interviews. As we noted in our discussion of researching users (in Chapter 4), you want to draw tools and methods for understanding users from different categories, allowing you to verify what you find from one with the results from another. Just relying on preferential data research could prove, as it did with the library website redesign, problematic.

The library team generated an initial prototype for testing the new website that was really complicated (Figure 6.4). There was too much on the proposed home page, including not one, but two search boxes, each in a different location. Each search actually accessed different data.

The overall thought behind the design of the new site was that students and faculty had different needs. Undergraduates, the team hypothesized, would want only basic information, such as library hours, and would also want only to search the library's general collection of books. Graduate students would have richer research demands but still have research requirements that would be less than

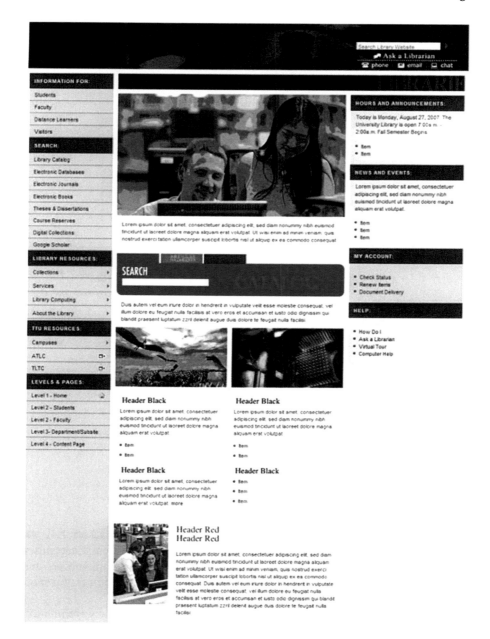

FIGURE 6.4   University library prototype website. (Courtesy of Dr. Brian Still.)

those of faculty. Consequently, the library team assumed that when users would access the site, although it might require some learning at the beginning, they eventually would know where their targeted search was located. To aid in parsing, the website's navigation included links named "Faculty" and "Students." When clicking on any of these, which the developers assumed would be the users' first entry point to content, images as well as labels on the page would change to match the user group; for example, images of faculty at work would appear on the faculty page.

To its credit, the library team knew it needed to test out its prototype, and Brian's lab carried out three usability tests, ultimately bringing in 45 users equally representative of faculty, graduate students, and undergraduate students. The results were not good. Not only did they speak to the need for better research of users, they also voiced as strongly the requirement that no product should ever be made for users that is not filtered based on needs, goals, and other pertinent factors. During testing, users were confused by the multiple search boxes, which was highly problematic because their first point of entry to complete tasks was to search, not to use any of the side navigation. Images for users were confusing or so large that they annoyed users. Graduate students and faculty often wanted the same things. In fact, undergraduate students also wanted the same things when they were required to research, and faculty wanted the same things as undergraduates when they wanted to know when the library was open or who to contact about a particular question. Much like the radio alarm clock for the hotel (see Chapter 3), the library's prototype website was too much of everything to be usable, given the situation for its intended users. Left navigation, middle navigation, right navigation, and multiple searches were all necessary, just like the clock requiring so many buttons to accommodate all the features of the prototype. But all this testing wouldn't have been necessary if the library developers had opted not to try to build into the site every possible feature. Less should be more.

Eventually, a lot of changes were made to correct and simplify the library's prototype site, but they were costly and time-consuming, primarily because they occurred after prototypes were made as well as after Brian and his lab were paid for their services. We are advocating that before you ever commit to what is your 1.0 prototype, take the time to filter full features into a manageable and reduced set of features.

This can be very simple. Pratt and Nunes (2012) suggest using a whiteboard (or a poster board to accommodate a group exercise). List out all of the suggested features and functionality of the product on the left side, and then write down all of the high-priority project goals and, just as important, user needs in the middle of the board. The idea is that you must do a test for each suggested feature. If you can filter it through goals and needs and it still has relevance, it moves to the right side of the board where you have reduced features and functionality.

So let's try to filter, again using EyeGuide Focus as our example product. In Figure 6.5, you'll first see all of the features that the EyeGuide team declared, based on its research up until this point, were necessary for Focus to be an effective and satisfying concussion detection system.

Once EyeGuide was satisfied with this unfiltered list, it set about determining if each feature met user goals and needs, client and organizational goals, and requirements of the situation.

Focus, as you remember, was created to be a 10-second concussion detection system that could be used quickly on the sidelines and provide an athletic trainer with a fast, reliable measurement that could be taken, along with other factors, to indicate the presence of a concussion. EyeGuide was working through the summer

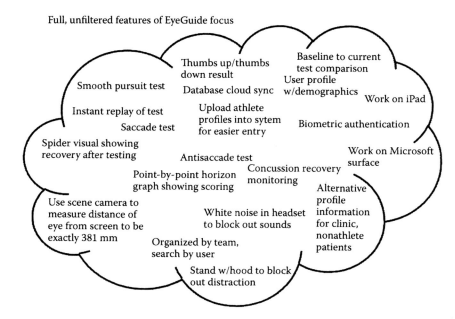

Full, unfiltered features of EyeGuide focus

Thumbs up/thumbs down result

Smooth pursuit test

Database cloud sync

Baseline to current test comparison

User profile w/demographics

Work on iPad

Instant replay of test

Saccade test

Upload athlete profiles into sytem for easier entry

Biometric authentication

Spider visual showing recovery after testing

Antisaccade test

Work on Microsoft surface

Point-by-point horizon graph showing scoring

Concussion recovery monitoring

Use scene camera to measure distance of eye from screen to be exactly 381 mm

White noise in headset to block out sounds

Alternative profile information for clinic, nonathlete patients

Organized by team, search by user

Stand w/hood to block out distraction

FIGURE 6.5   Focus features before UCD filtering. (Courtesy of EyeGuide.)

months of 2015 to prepare Focus for alpha or trial testing with a select high school football team, so it faced a time crunch. That immediately meant that any features that might require more development, even if they could enhance the product, would likely need to be put off for later implementation, especially if they weren't necessary for Focus to accomplish the most important goal for the user and the organization, which was to detect a concussion and report on it. Consequently, features such as an integrated biometric authentication or a form to allow patient data at a clinic were ruled out right away as being unnecessary for a 1.0 Focus. They didn't meet organizational goals as well as time (situational) demands.

Although the ultimate goal for Focus was to work on multiple platforms, the athletic trainers that would participate in the initial study preferred iPads as the platform tool. As great as it would be for Focus to eventually work on a Microsoft Surface or even an Android tablet, those features were eliminated because initial users didn't need them to accomplish their goals.

Athletic trainers also told EyeGuide developers they had no plans to test on the sidelines. Their protocol was to take athletes directly to the locker room and test them in privacy, away from the crowd noise and other distractions. This meant that features such as a stand with a hood, something EyeGuide thought necessary to shield athletes being testing while on the sidelines, wasn't necessary and was filtered out as was the proposed feature that would pipe white noise through the headset earphones to eliminate any possible distracting sounds.

Perhaps the longest discussion, with frankly a lot of arguing, ensued when determining which test features were required to collect the data and present it to make Focus a viable 1.0 product. There were three suggested tests: smooth pursuit, saccade, and antisaccade. Each one on its own, as previous research had shown, could indicate a possible concussion. However, all three taken together were much more reliable. But EyeGuide developers and key stakeholders, including the CEO, debated if it was necessary to have all the tests when one might be good enough. One test meant less time to test, less setup for the athletic trainer, and less interpretation of results. However, one test also could mean that athletic trainers might not be convinced.

Ultimately, for a 1.0 test trial, Focus eventually included just one test, a smooth pursuit test that required users to follow a small, white reticle on the iPad screen as it moved deliberately tracing an invisible lazy 8 pattern. The literature better supported smooth pursuit as a standalone test, it was easier for the developers to implement and score, and feedback from the athletic trainers indicated they understood what it meant more easily than the other tests when they viewed different results from each one.

Finally, even greater debate raged on, and still goes on to this day, about how best to display actionable, believable results to athletic trainers so they can use it to make a fast, confident determination about an athlete's condition. Focus interviewed athletic trainers, experimented with various visuals for displaying the results, and created paper prototypes as well as opportunities for users to interact with them and give feedback on what they wanted to see, even before Focus was an actual prototype. A few athletic trainers wanted side-by-side charts that showed them different views of the data that they could analyze and, along with other factors such as athlete complaints about dizziness or lack of memory, make a decision.

But most athletic trainers wanted something akin to a concussion thermometer. They knew how, qualitatively, to assess a concussion. What they wanted was a quantifiable, quick confirmation of their assessment. EyeGuide, as discussed in Chapter 5, wanted to separate itself from competitors by being not only reliable, but also fast and mobile. Longer testing and larger amounts of data feedback would run counter to those goals. For 1.0 Focus, EyeGuide matched up the different result display features to user goals, organizational goals, and situational requirements and, in that process, determined that a single result visual (Figure 6.6) with a thumbs up or thumbs down image, a statement identifying the nature of the score (ranging from severely impaired to very superior), and a static image showing the athlete's eye movement overlaid on the optimal lazy 8 path, was the only visual result, after filtering, that was needed.

The end result of all the filtering of the full set of EyeGuide Focus features was a reduced set of features (Figure 6.7).

You may eliminate only one or two features. EyeGuide elected to get rid of 11 of its planned features for Focus, leaving only nine (less than 50% remaining). Was that a lot? Maybe, but it really depends on user goals, organizational goals, and situational requirements. You can always restore features that you previously cut after testing when users ask for it, or later in 2.0, 3.0 versions, and so on. Just remember, not filtering means you run the risk of giving too much to the users for them to handle, just as it also can mean you spend more than you should for things you don't need, run behind schedule, and even risk failing.

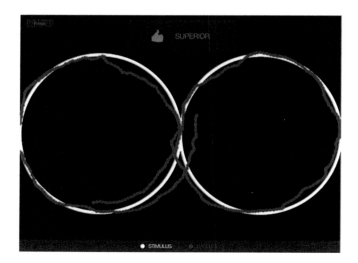

FIGURE 6.6 Focus concussion test results visualization. (Courtesy of EyeGuide.)

Also, ironically, by choosing not to filter, you are in fact, filtering. In doing so, you aren't thinking about user goals and needs, the situation, or even your organizational or client demands; instead, you are making a call about what you think is best, which means you've filtered yourself. That isn't UCD—that is DCD (designer-centered design).

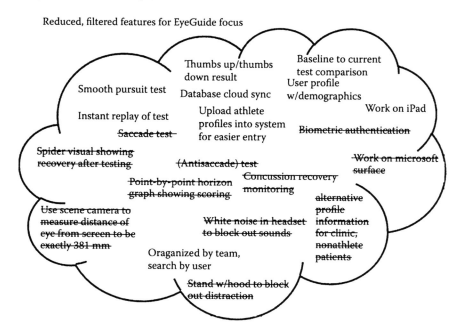

FIGURE 6.7 Focus features after filtering. (Courtesy of EyeGuide.)

6.2   TALKING TO CLIENTS

Before we finish this chapter and the "balance needs" part of the RABBIT process, let's devote a little time to how to talk with and include client feedback. One of the best ways to ensure harmony and avoid major disappointments and scrapped projects is to sit down with clients and get a clear idea of their expectations and purposes for the project. You may also need to describe the UCD process to them—especially if they are unfamiliar with it—so that they can know what to expect from you or your team and the way iterative discovery and design works.

In UCD, ideally, designers regularly touch base with clients multiple times throughout the development of the product to ensure that it is going in a direction that clients are willing to explore. They can also help you define priorities for the design, such as functions or features that are primary versus those that can be cut if need be. In addition to discussing goals, you'll also benefit from sharing concept designs and prototypes with your clients so they can, in the least, warn you away from approaches or options that they can't afford or support. Although they may have hired you to be their designer, they are beholden to spending limits, market and board influences, and company policies.

Here are some questions you might ask during your meetings with clients:

- What are your initial ideas for user needs that you want to appeal to or fill? (i.e., What are you trying to achieve?)
- What design concepts have you already started developing?
- What features or functions do you expect to see in our product design?
- How much time do you expect the project to take or will it be ongoing?
- How soon do you expect to see a prototype?
- What are our resources, in terms of money, staff, materials, space, and equipment?
- What specific markets or products are you trying to compete with? Who are your competitors?
- What niches are you hoping to fill with new or revised designs?
- What are your proposal and reporting standards?
- What policies should we be aware of before we make recommendations for designs? In other words, what are some of the affordances or constraints on product development?
- What data resources can you make available to us, such as user feedback data, consumer purchase patterns, previous user testing data, etc.?
- How often can we meet with you to discuss design development and collaborate on ideas for new iterations?

All this said, you don't want to rely too heavily on client input. They may have a very different idea of what users need or want, which might contrast with what real users want. In addition, you've been chosen as the designer for a reason, hopefully, and that's because you come up with great ideas and insights regarding the users' experiences. Conveniently, many of the questions listed above can be asked and answered via email.

Meetings with clients should probably focus on reviewing concept or prototype designs and getting their feedback. But in the end, you'll just want to make sure you are prioritizing their needs along with users' needs.

TAKEAWAYS

1. Just because you have the money and ability to add a feature or function—even if the user wants it—doesn't mean you should add it. Your goal isn't to give users everything and overwhelm them with features; alternatively, it is to give users what they want and need within the particular use situation to be effective and work efficiently so they derive some pleasure or satisfaction from the experience.
2. Balancing needs is about prioritization and filtering; you have to give priority to what will make the product work best for users and, at the same time, achieve your organization or design goals.
3. Use triangulation to filter down your full set of possible product features into a reduced set that meets user and organizational goals.
4. Filter features before you commit to a 1.0 prototype. You can always restore a feature that you may have previously cut if users ask for it in the future.
5. If you don't filter features up front, you run the risk of giving too much for users to handle, spending more than you should for things you don't need, running behind schedule, and having a failed product.
6. If you bombard users with too many options to satisfy what you perceive to be your goals or theirs, you will overwhelm users so much that they are effectively paralyzed into inaction.

DISCUSSION QUESTIONS

1. Why do you think so many products have so many useless features? Where do you think the breakdown in the UCD process occurs?
2. Do companies *have* to add extra features (even if they know users probably won't use them) to be competitive or differentiate their products from other companies?
3. How do you prioritize product features with organizational or client goals when there is disagreement about which features should be priorities? For example, what if you know users prefer a particular feature, but the paying client wants to use a different feature because they think it's "cooler?" How can you make a compelling case about which feature is truly a priority?
4. What potential points of resistance or confusion might a client have about the UCD process? How would you address these?

REFERENCES

Fox, Z. "The Average Smartphone User Downloads 25 Apps." *Mashable*. September 6, 2013, accessed February 21, 2016. http://mashable.com/2013/09/05/most-apps-download -countries/#s89rqffw_ZqH.

Jones, B. "Understanding the F-Layout in Web Design." *Envatotuts+*. March 7, 2012, accessed February 1, 2016. http://webdesign.tutsplus.com/articles/understanding-the-f-layout-in -web-design—webdesign-687.

Lipsman, A. "How the Power of Habit Drives Mobile App Usage." *comScore*. September 23, 2015, accessed February 21, 2016. https://www.comscore.com/Insights/Blog/How -the-Power-of-Habit-Drives-Mobile-App-Usage/.

Nielsen, J. "F-Shaped Pattern for Reading Web Content." *Nielsen Norman Group*. April 17, 2006, accessed February 21, 2016. https://www.nngroup.com/articles/f-shaped-pattern -reading-web-content/.

Pratt, A., and J. Nunes. *Interactive Design: An Introduction to the Theory and Application of User-Centered Design*. Beverly: Rockport, 2012.

Schwartz, B. *The Paradox of Choice: Why More Is Less*. New York: HarperCollins, 2004.

7 Build Out an Operative Image

If you've come to this point in the book expecting details on what colors to select, interactive features to use, or layouts to put in place that work best to design a UCD product, you'll likely be disappointed. Certainly, color, typography, interactivity, and layout all are important considerations for making an effective product. But this is a book about UCD. The colors you choose and the way you organize elements of an interface or employ particular features that promote interactivity are all dependent upon the needs and wants of users. In addition, situational demands placed on where users will use the product, governed, as you've learned, by a variety of factors including your organizational or client goals, product materials, environment, and so on, should also impact your product's construction. Our emphasis, therefore, as we've noted before, is on designing a product from a user-centered perspective.

What colors should I choose? Obviously, you will want to rely on best practices, but as part of a UCD process, the best way to choose the right colors is to involve user feedback: Research your users, look at what competitors are doing, consider situational factors, and ultimately build a prototype that you can then provide to your users to get their feedback. That feedback will tell you about the effectiveness of your product's colors, layout, and other features. We are designing for users and putting users at the center of the design process. We don't choose colors, layouts, or features without first researching what will work best for our intended users and then allowing those users to test what we've created for them.

To be able to give your users something to test early in the process, you're going to need to create some version of your product that users can engage with. It can have a low fidelity (lo-fi)—remember Jeff Hawkins' wooden Palm Pilot or Brian's white cardboard box that represented the radio alarm clock. The key is moving from a vision state, which according to Lowgren and Stolterman (2004) "is only in the mind of the designer" (18). Because a vision is "sketchy and diffuse," users cannot interact with it (18). What's needed is an "operative image," they continue, or the "first externalization of the vision" (19). This operative image, Lowgren and Stolterman write, can be "a diffuse image and is usually captured in simple sketches" (19). However, because it is something that takes the form of the product, it can be "put to the test" by users (19). As a result, "It will be challenged by new conditions, restrictions, demands, and possibilities. The image becomes increasingly detailed and complete" (19), governed in this transformation by user needs and wants as realized through their feedback.

In this chapter, our focus is to show you how to render the knowledge you've gained into an operative image or prototype that has the look and feel of your eventual finished product, allowing you to test its effectiveness with its intended users. What you create can be a version of the entire product, or just parts of it, such as the interface. Its fidelity, or how much it is like the final product, can vary. We often tell our students that their prototypes aren't ducks, but they should look enough like ducks, and act enough like ducks, and maybe even quack like ducks to make users think they are seeing ducks.

7.1   USING THE MVP

At the earliest stage of design, you can create what Gothelf (2013) calls a minimum viable product (MVP). The MVP allows designers to "build–measure–learn" quickly (Gothelf 2013, 7). Faced with the challenge of trying to get feedback from users in environments in which developers code and release quickly (called agile development), designers came up with the MVP as a constructive way to create a prototype as fast as possible, have users use it, then make changes to it and test it again. Brian wrote about this process in a column for *Intercom* in 2014:

> Rather than wait to engineer a complete design, or evaluate even a completed, low-fidelity prototype using any number of analytical methods, Lean UX encourages not just UX professionals but also developers themselves to work together, moving as quickly as possible, to generate business solutions. Rather than think about building a finished product to get feedback, developers should try to figure out what users need, hypothesize on what the best way is to deliver on that, and then test it by having real users interact with the prototype. (Still 2014, 28)

The MVP you create can take many different forms. You also don't need to sit down with users and carry out face-to-face testing. The idea behind the MVP is to put something together quickly that resembles your intended product, then release it and gather feedback. What you create doesn't have to be real. For example, Eric Ries (2011), the founder of Zappos, writes about how he used an MVP technique called the "Wizard of Oz" to see if people would really buy shoes online. The founder created a web page with photographs of shoes and their prices and then made it seem as though users could select and purchase the shoes. Once he learned from user interest that they would, he actually went to a store and bought the shoes to fulfill the orders manually. What people thought was magic was really just a guy behind the curtain making it seem like there was magic. However, this MVP confirmed to the Zappos founder that people would be willing to go online and buy shoes, and now Zappos is the world's largest online retailer of shoes.

The MVP is a fast way to test out a hypothesis about your product. Perhaps you have reached this stage of the RABBIT process and you still don't feel like you know enough about your users to create a full prototype of the product. For instance, you might have different options for content, color, or other features to choose from, but you aren't confident enough to pick which ones should go into the product. Use an MVP. Rob Kelly (2011) tells the story of Cars Direct, which wanted to sell cars online in 1998. Instead of building out a more finished prototype of an online cars website, Cars Direct started simpler. Like Zappos, Cars Direct just wanted to know if someone would actually buy what it wanted to sell online. Would someone use a credit card to buy a car online? To learn the answer, Cars Direct posted a few images of cars with prices and purchasing options online. The next day, according to Kelly, users had bought four cars. The MVP that Cars Direct created not only answered the question of what users would be willing to do, but it also told Cars Direct, according to Kelly, that the rest of the website built up around selling cars didn't have to be feature rich. Users wanted to buy cars, so Kelly realized the website design should make the process as straightforward and easy as possible, eliminating or pushing into the background other things that didn't support this.

7.1.1 CREATING AN MVP

Gothelf (2013) suggests the following guidelines for creating an MVP:

- Be clear and concise.
- Prioritize ruthlessly.
- Stay agile.
- Measure behavior.
- Use a call to action.
- Be functional.
- Integrate into existing analytics.
- Be consistent with the rest of the application (57–58).

Here's an example of how these guidelines are put into action. As you know by now, EyeGuide, after its failures with Assist, set about creating a new product relying on UCD. That product, a 10-second concussion detection system called Focus, was intended primarily for athletic trainers to use on the sidelines or in locker rooms to detect concussions and then monitor the recovery of athletes who had been diagnosed with concussions. But an early stage question for EyeGuide was, "Would athletic trainers want to use a product like Focus, which used eye-tracking technology and an iPad for results display?" Although scientific literature supported the fact that eye movement and impairment of eye movement signaled accurately the possible presence of concussion, most tools athletic trainers used for concussion detection involved other methods quite disparate from what Focus was conceived to be. Eventually, Focus was built out into a more fully developed paper prototype and then a wireframe, tested in both instances by athletic trainers. We'll see examples of these operative images soon. But before reaching these points of development, EyeGuide wanted to know if what they were considering was viable enough to progress from vision to operative image.

As we noted previously, the RABBIT process isn't necessarily linear. You can build out an operative image even in the midst of researching users or carrying out other types of knowledge gathering if it aids in determining user needs and wants. The MVP offers an ideal means of doing this because it can be deployed so quickly with little expense. EyeGuide felt relatively confident about its understanding of athletic trainers and where they would use Focus, but it had a big question about whether athletic trainers would consider the product a viable tool. In technology-centered design, we build a product and expect people to adapt to and use it; conversely, in UCD, we build a product that meets user needs and wants. Our first inclination, in fact, is to answer the question, "Do users need and want it?" The MVP facilitates well answering such a question or derivatives of it.

EyeGuide took advantage of an already scheduled meeting that included a few athletic trainers. Its developers created an MVP that had just three screens, all designed to display on an iPad:

- One (Figure 7.1) showed the pupil lock, demonstrating the accuracy of the eye tracking technology and how the testing would work.
- Another (Figure 7.2) showed athletic trainers what would happen once the eye tracking was in place and the test was completed.
- A third screen (Figure 7.3) showed results, similar to the screen demonstrated in Chapter 6, that athletic trainers would receive after a test was completed.

FIGURE 7.1   EyeGuide Focus pupil lock demonstration (screen 1 of 3 of MVP). (Courtesy of EyeGuide.)

FIGURE 7.2   EyeGuide Focus test complete confirmation (screen 2 of 3 of MVP). (Courtesy of EyeGuide.)

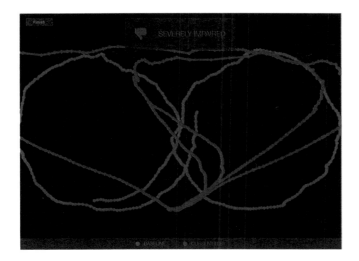

FIGURE 7.3 EyeGuide Focus test results screen (screen 3 of 3 of MVP). (Courtesy of EyeGuide.)

As Gothelf's guidance (2013) prescribes, the EyeGuide MVP was clear, focused on only the most important issue at that moment, built in PowerPoint so it could be easily modified, functional (the eye captured in Figure 7.1 was a live image of an actual user's eye), and similar to other EyeGuide products, and with athletic trainers present in the room, it was also set up to gather immediate, actionable feedback from the target audience.

What EyeGuide learned was that the athletic trainers present weren't averse to the technology, primarily because of the comfort level they had with the iPad. Athletic trainers also didn't object to eye tracking as a means of detecting concussion because they had some knowledge of the literature about eye movement and impairment. They were, however, concerned about ease of use of the technology, and they also indicated that they would likely be more interested in Focus as a recovery tool because, they argued, a wide range of factors went into determining a concussion. Still, the fast results of Focus, as the MVP suggested were possible, had value, they thought, in confirming a suspected concussion. Certainly, the use of the MVP, costing almost nothing and making use of existing technology, gave confidence to the EyeGuide team that it would be worth the effort to pursue building out more refined prototypes of the Focus concussion detection system.

You, too, should consider using the MVP when you are early in development or when you have a question related to users' needs and want to answer it as quickly and decisively as possible. An MVP can be almost anything as long as you know what it is supposed to do and how its design, whatever shape it takes, serves to mimic something about the product's content, features, or purpose.

7.2   PAPER PROTOTYPING

Carolyn Snyder, who literally wrote the book on paper prototyping (2003), calls it a design method that "lets you get maximum feedback for minimum effort. After a few usability tests with a paper prototype, you'll have confidence that you're implementing the right thing" (Snyder 2001, par. 1). When we say paper, we mean low fidelity. Obviously, the lowest fidelity for a website, for example, would likely be a paper-based rendering of it. But you also could, as we have many times, use a whiteboard with dry-erase markers to sketch out a paper prototype or low-fidelity version of a website.

The point of paper prototyping, especially as a design tool, is that it allows you to involve users to interact with a simulation of the product early on in the process. The drawback is that the paper prototype isn't really functional like a website or any other high-fidelity product like software, hardware, or a mobile app that it simulates. Consequently, you cannot really gather more definitive, quantitative measures of the prototype's effectiveness, such as counting how many mouse clicks it takes to find something, how long it takes to complete a task, or how long users take between steps (called dwell time) when trying to do something.

But you make up for all of this by encouraging users, through their use and what they tell you as they work, to give feedback that illuminates what are often deeper issues related to the intuitiveness of the product you're trying to create. As Rettig (1994) argues, paper prototyping promotes quality enhancement:

> If quality is partially a function of the number of iterations and refinements a design undergoes before it hits the street, lo-fi prototyping is a technique that can dramatically increase quality. It is fast, brings results early in development (when it is relatively cheap to make changes), and allows a team to try far more ideas than they could with high-fidelity prototypes. Lo-fi prototyping helps you apply Fudd's first law of creativity: "To get a good idea, get lots of ideas." (22)

Remember, we've talked a lot about mental modeling or how users view their world and how products need to be created to fit into that cognitive vision. Paper prototyping enables users to be proactive producers and, therefore, key contributors to the design process; as Still and Morris (2010) write, it "allows users to fully express their conceptualizations and, thus, reveal their mental models of a proposed design" (144).

Although the paper prototype mimics the final product, it isn't coded yet, so "designers are willing to make changes," according to Still and Morris (2010), "and users are more comfortable suggesting them. The result is that real users have a substantial say, beginning early on, about the products designed for them, and this almost always translates to better usability, fostering an environment where users are a key part of the design process" (144). There's a tradeoff, clearly, because the paper prototype isn't a high-fidelity product. But precisely because it is low fidelity, users are less reluctant to modify it. In fact, the less it feels like a finished product, the more likely they are to interact with it and feel comfortable modifying it, and this is exactly the contribution you want at this stage.

"I cannot emphasize enough," Medero (2007) writes, "how important the inclusive quality of paper can be. Though some people shy away from paper prototypes because they feel they will not be taken seriously, I argue that many people are intimidated by a formal, highly technical design process and that the less 'professional' nature of paper prototyping is a great way to lighten the mood and engage a more diverse group" (sec. 1, par. 2). You want a version of the product that looks and feels like the product, but you also want to encourage users to use it and, through that contribution, help you design a better version of the product.

Your paper prototype can be designed so that users can pretend to click buttons or interact with dynamic parts of the product in order to complete tasks you've given them to perform. For example, you may have a paper prototype of a pizza restaurant's website home screen. You could ask users to find the location of the nearest pizza restaurant. Users, in trying out the paper prototype, would "pretend" click on the locations link in the menu or the search box. You then could provide users another piece of paper showing users the results of their actions, such as the location's main page or a search box that would allow users to enter terms.

Why do this with paper? What if users don't find a menu item that makes sense as the direct path to locations? If you conduct a high-fidelity test, users are going to try to use the website. They may fail to find something, but why? Well, it could be because there was no button that was labeled locations, but you don't know that. You can only assume. However, the paper prototype, because of its fidelity, would allow users to change a button to read "Locations" or even add a button for locations if one didn't exist. As a result, paper prototyping encourages modification, early on, which saves on costs because you're able to make changes before the front and back ends of the website have already been coded.

If users want to do something that you haven't built, such as attempting to navigate to a page you haven't created for pizza coupons, you know they are thinking about the product differently than you expected. You can, in such situations, either return them back the paper prototype to ask them to use what you've made, or you can give them a blank piece of paper and tell them, "That page doesn't exist yet. Would you mind sketching out what you think should appear there?"

7.2.1 Creating a Paper Prototype

7.2.1.1 First: Build Your Paper Prototype

The prototype can have a lot of detail, looking almost like a paper version of your final product, or it can have elements missing, on purpose, because you want users to help you add those. You can create the prototype to match the approximate size of the final product, such as the Hawkins Palm Pilot wooden block. You also can get nicer paper of card stock quality or use regular notebook paper.

Feel free to hand-draw the prototype's features or use Microsoft Word, PowerPoint, or any other tool you're comfortable with to create a more finished prototype interface that you then print off for use. You also can choose to keep the prototype in black and white or add colors and even images to increase fidelity. Rettig (1994) writes that "the photocopier can be a Lo-Fi architect's best friend. Make yourself lots of blank windows, dialogs, buttons and fields, in several sizes" (24). Create extra copies of your prototype so that each user has a clean version to use.

You are creating a prototype that looks as much like the duck as possible, given your goals, so users can interact with it. But you don't want it to be the actual duck. Instead, you want to create the prototype quickly, based on what you know, and to quickly learn what you don't know so you can quickly fix it to make it better; further, you want to do this as many times as possible as soon as possible. Remember UCD Commandments I and II: Thou must involve users early, and thou must involve users often.

7.2.1.2 Second: Define Goals for Your Paper Prototype

You may want users to tell you everything, and so you start with literally blank paper. It could be that you already have a preexisting template in mind, so you design that and then require users to react to it. Whatever you decide to do, you need to have a purpose for the test. Why are you asking users to interact with this paper proto-type? Given where you are in the RABBIT process, you have a decent sense of your product's users, and you have a filtered features list for your product. What do you want to learn from users interacting with a paper prototype that will enable you to move forward to the next stage of development?

Create scenarios for users to operate inside of, representative of the actual use environment, as they offer feedback. Brian's team wanted to know what about the hotel alarm clock was problematic. So, as Box 7.1 shows, they created a scenario that described the situation and gave the users a clear task to perform to generate feed-back based on an actual use situation.

It's important to have a scenario or scenarios, a story that requires the users in question to carry out tasks. In some cases, you can pretend they're interacting with an actual user interface. When that happens, you tell them to treat the paper just like the user interface they're using (a website or app, for example). When they need

---

**BOX 7.1   PAPER PROTOTYPE SCENARIO
(COURTESY OF EYEGUIDE)**

You're a business traveler who has to stay in a lot of hotels. One of the things that you need, like a lot of travelers, is a really good, reliable alarm clock, one that provides you with the functionality you need, can play music as necessary, perhaps even do other things. We're in the early stages of developing a radio/alarm clock for a major hotel chain, one that will eventually be used by count-less travelers just like you. If you could design a radio alarm clock so that it had the features you wanted, and those features were easy to use, and the language to describe them was easy to use, what would your clock look like? You'll see the paper box in front of you. It's shaped like a radio alarm clock. Some of the things you see on it, like the on/off switch and volume control, have to be where they are. The rest is up to you. Build your alarm clock with the buttons we've provided. If you don't see a button, create your own.

to click, they tell the computer (played by someone on the team), and the computer makes the navigation happen.

7.2.1.3 Third: Gather Materials for User Interaction

Plan to provide materials for users to interact with; these materials could mean any of the following, depending on what you mean by paper or low fidelity:

- Markers, pencils, and pens (of different colors).
- Different-colored paper or Post-It notes—these can be cut out and taped to indicate, for example, menu items, or users can use Post-It notes to do the same thing, writing content of say a navigational button on a Post-It and applying it to the page.
- Other items, such as photos, logos, etc., that you know you want or need to have appear on the product—users can then select these and place them where they feel the best place is for them to go.

How many and what variety of materials you require depends on how much control users will have in modifying the prototype's design.

7.3   EXAMPLE PAPER PROTOTYPE

Let's now take a look at an example of a paper prototype. After creating the MVP and gathering feedback about athletic trainers, EyeGuide chose next to create a paper prototype, which had fuller features than the MVP. The MVP was focused primarily on determining if (1) athletic trainers would need or want an eye tracking-based concussion detection system, and (2) they would accept a particular visual (already shown) for displaying test results. Having learned that they would accept the system and generating additional knowledge (through a design studio exercise we'll examine shortly) about what athletic trainers expected to see in a concussion detection interface, EyeGuide wanted feedback (including design input) from athletic trainers about a concussion detection system that flowed from athlete profile creation, to testing, to results display. Namely, developers, led by Miriam Armstrong and Megan Olson (2015), were most interested in how athletic trainers responded to Focus as a concussion recovery monitoring tool. Athletic trainers had shown a willingness to accept Focus as a concussion detection tool, but what features and navigation would be required for them to want to use Focus to keep track of athletes' recoveries from concussions?

The paper prototype EyeGuide developed was hand-drawn but organized in a layout that would work for an iPad touchscreen. Additionally, navigational elements were included, such as a player profile search bar on the left as well as a main content area that displayed an individual player's information. Following from this were two other hand-drawn, simulated screens: One showed the player's total test results after injury, monitored by test in a line graph, and the other showed symptoms the player continued to report (Figures 7.4 through 7.6).

To collect verbal and performance (say and do) feedback that achieved EyeGuide's goals, users were given the scenarios and tasks outlined in Box 7.2.

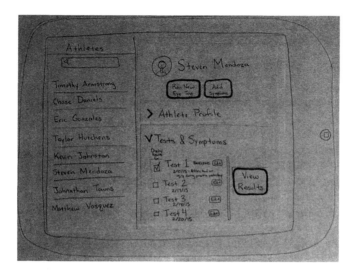

FIGURE 7.4   Focus paper prototype main athlete profile screen. (Courtesy of EyeGuide.)

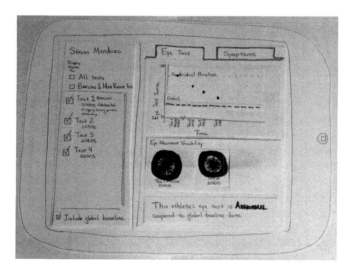

FIGURE 7.5   Focus paper prototype athlete test results history screen. (Courtesy of EyeGuide.)

The results from users interacting with the Focus paper prototype allowed EyeGuide to make necessary changes. For one, athletic trainers responded quite inconsistently about how recovery should be displayed; based on this feedback as well as other issues, including development timeline, EyeGuide decided to forego including concussion recovery monitoring in the 1.0 version of Focus. Additionally, EyeGuide learned there were some elements missing in the individual athlete's demographic profile that athletic trainers expected. Those were included in the next round of development.

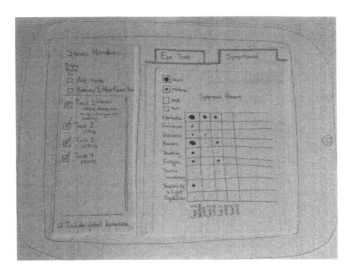

FIGURE 7.6   Focus paper prototype athlete symptoms screen. (Courtesy of EyeGuide.)

BOX 7.2   FOCUS PAPER PROTOTYPE
SCENARIOS (COURTESY OF EYEGUIDE)

You can see how the program has recorded information about where the athlete was looking in relation to where he or she was supposed to look. Athletes without concussions will generally be better at following the white dot than those without concussions. Their eye movements will have less variability than those with concussions.

So far, there is no way to look at how well the athlete did on the eye-tracker test other than watching the replay video. EyeGuide has started designing other ways to display the eye-tracker information. You'll help us get an idea of what is good about the design and whether it has any problems.

As I've mentioned, the design is in the very early stages. Right now, we just have a version that is made out of paper. You can still use it like you would an iPad app; Meghan will be playing computer. When you press a button, she will update the screen.

We're going to give you a few tasks to perform on this application. Again, we are not testing you. The tasks are designed to test the current design.

The iPad program in front of you has the information of athlete Taylor Hutchens pulled up. Taylor was diagnosed with a concussion last week. Since then, you have given Taylor the eye-tracking test several times. The EyeGuide Focus app has all of the eye-tracking tests stored, including a preinjury baseline test the athlete took at the beginning of the school year. Additionally, the global baseline data, the averages for all athletes, are stored in the system.

Task 1: You remember that Test 1 was the baseline test the athlete took at the beginning of the school year. Indicate to the EyeGuide Focus program that Test 1 is the baseline. When you have finished, please say to me, "I've finished."

Task 2: Ultimately, you would like to decide today whether Taylor should continue rehabilitating or whether you can recommend that he see a physician or resume play. Select the tests you would like to compare in order to make that decision. When you have finished, please say to me, "I've finished."

Task 3: With the information stored in the system, decide whether Taylor should continue rehabilitating or whether you can recommend that he see a physician and resume play. When you have made your decision, please tell me that he should either continue rehab or see a physician.

The paper prototype can be more comprehensive, in our experience, than the MVP. The MVP answers the hypothetical question quickly and effectively. But the paper prototype offers a fuller opportunity for feedback from users about all aspects of the simulated product. Its low fidelity does limit it, but the interactivity it promotes and the encouragement it gives users to act like designers makes the loss of fidelity worth it, especially early in the design process.

7.4   MEDIUM-FIDELITY WIREFRAME

Once you are further into design, perhaps past the paper prototype when you've captured user feedback and caught, in the process, the major cognitive obstacles that stand in the way of them adopting and enjoying the product you are making for them, you can consider implementing another design prototype of higher fidelity: the wireframe.

Unlike a lo-fi paper prototype not rendered into a form that can be interacted with on the product's eventual platform (e.g., website on computer, mobile app on a mobile device), the wireframe should be delivered for user feedback in a more realistic environment.

But how much of the duck it acts and looks like depends on where you are in your design process. It could be that you have learned enough about user needs and wants and are ready to move on from what Garrett (2011) describes as the strategy, scope, and structure layers. However, you're not ready to mock up a higher fidelity wireframe with content, images, and other features; therefore, your wireframe will reflect the skeleton of the product, showing the layout of the interface, for example, and the placement of key features and other elements. There's no need to use color unless it is absolutely necessary for this medium-fidelity wireframe. You don't necessarily need to have labels on your content blocks. You're trying to get a sense from users if they can use—or even want to use—the product as you've laid it out. At this stage, you're attempting to give users a look at what the product may ultimately present to them. It also informs designers of the plans for the product so that later, after user feedback, they can craft a higher fidelity wireframe.

Figure 7.7 is a Focus wireframe that places emphasis on the skeleton of the product interface.

You can see that, at this stage of the product's design, Focus developers had conceived of an interface that would allow athletic trainers to pull up an individual athlete from a roster on the left. Even this wireframe was tested with user feedback. The wireframe was, as the figure shows, a simple black and white image (in PNG format) displayed on an iPad.

The gestural features normally found in applications designed for a smart screen device, like an iPad, didn't work in this medium-fidelity wireframe. Still, during its testing, users were encouraged to use their fingers to scroll up and down to find, for example, a player from the roster, or to push buttons, such as run test, to simulate a concussion test. Essentially, this medium-fidelity wireframe served as a functional prototype for EyeGuide. Users were able to understand how things would work and could comment on the flow of actions and the layout of elements, but they weren't distracted too much by colors, interactivity, or even content because that wasn't the point of the test.

You should design and test, design and test, improving the product each time, and you should do it as much as possible. From MVP to paper to medium fidelity and possibly with different iterations within each level of fidelity, focusing on different aspects of the product's design, you want user feedback to inform your design. You don't build and guess. You build and test with users. It may be that, after constructing and getting feedback on a wireframe similar to the Focus example, you choose next to add in text, perhaps also color, so that you can see how users feel about the

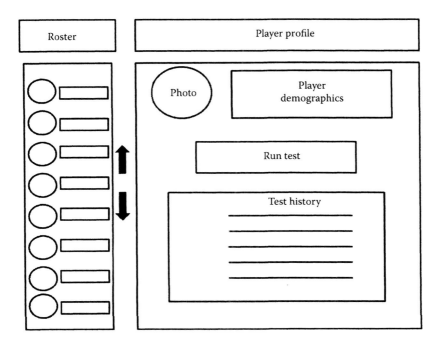

FIGURE 7.7 Focus wireframe. (Courtesy of EyeGuide.)

layout of the interface at a little higher fidelity. It still wouldn't be fully interactive, but before committing to color and content in a functional prototype, you could test one more time with users.

7.5 HIGHER FIDELITY PROTOTYPE

The higher fidelity (hi-fi) prototype looks and acts a lot more like the duck. In fact, to users, it might appear to be the duck. Buttons can be clicked on a computer or gestures swiped on a mobile device; search boxes can be used and forms completed. For the most part, a lot of the product's interface will appear completed, rendering a prototype that feels and looks like the final product. However, the back end of the product will likely not be complete. No live data will be flowing to the hi-fi's interface. Buttons or swipes may work but will only send users to predefined locations that exist. If users try to do something requiring a smarter back end or if they head in a direction that doesn't exist, then the hi-fi's true nature will appear.

The hi-fi is really an in-between prototype and for a reason. You've taken an informed leap to get to here. You've done a lot of research, and you may have also already built a couple of lower fidelity prototypes and tested them with user feedback. But you're not yet ready to expend the time and money required to build out a final product and go live. You want more feedback. You're confident you've discovered major issues—the types of problems Garrett (2011) tells us lurk at the scope, strategy, and structure layers. You've even tested the skeleton of the product with your medium-fidelity wireframe. Now you want to see if you understand what users need and want in a more fully developed version of your product, something that lets users interact with the skeleton and surface layers of the product.

The hi-fi offers an ideal way to gather this kind of feedback, quickly and cheaply. In addition, user feedback about what users see, do, and say when interacting with the hi-fi still comes early enough that changes can be made to improve the product. Whereas the MVP helps us answer a single hypothetical quickly, and the paper prototype lets us give users a chance to weigh in, actively, on designing a product that works for them in a nonintimidating environment, the hi-fi allows us to test if everything we've learned about user needs and wants has been adequately addressed.

We recently learned that more NFL quarterbacks (Smith 2015) are increasingly using virtual reality headsets to prepare for the defenses they will see in the next week's game. Nothing is quite like the real thing, but watching a film of the game or drawing up a play on a whiteboard to counter a defense is not as real as wearing a virtual reality headset that allows a quarterback to visually, virtually at least, preview the other team's defense. Think of the hi-fi in the same way. Others are also turning to this in-between, augmented or virtual reality training, including the military and law enforcement. You want to know if what you've done works or will work. You've prepared in every way possible. But just before you get into the real game, or, in our case, before we start spending a lot of money going into production or marketing to the world that we have the latest and greatest next-generation product, we'd like to try out a more refined version of it just in case.

Thanks to 3-D robotics printing, EyeGuide can effectively create hi-fi prototypes of all of its headsets and other hardware relatively quickly and affordably. We spend more time looking at how EyeGuide carries out RABBIT UCD for hardware in a later chapter. However, we want to note here that hi-fi prototyping for hardware works basically the same way as it does for software, websites, mobile apps, and even printed UCD products. The hi-fi headset represents a lot of research, including talking to users, modeling competitor products, and designing 3-D versions for display on the computer. At some point, however, we can only learn if the product works once users wear a headset and use it. The hardware hi-fi allows for this to happen without having to spend tens, if not hundreds, of thousands of dollars on plastics molding as well as other processes, including electronics.

For most of you, your hi-fi tools will not be 3-D printers. Rather, you will make use of applications such as Pop, Balsamic, or Axure to craft and deploy. All of these and others, some that you have to pay to use and some that are free, are mostly easy to learn and implement.

7.5.1 FOCUS HI-FI PROTOTYPE

Figures 7.8 through 7.20 represent screenshots in the hi-fi Focus prototype.

EyeGuide developers actually printed these screens initially to carry out a walkthrough of Focus's flow from adding a player to carrying out a concussion test. In other words, they reduced the fidelity from an onscreen, interactive prototype back to a more refined but nevertheless low-fidelity paper prototype, so they could be sure, based on user feedback, that how Focus guided users from start to finish made sense. As we've noted, you should be flexible, based on your goals, for the fidelity of the prototype you use. If it helps you understand how users feel about something you've developed, either the entire product or just one aspect of it, then you are encouraged in a RABBIT UCD framework to do what's required to learn what needs to be learned.

FIGURE 7.8 Hi-Fi Focus prototype screen. (Courtesy of EyeGuide.)

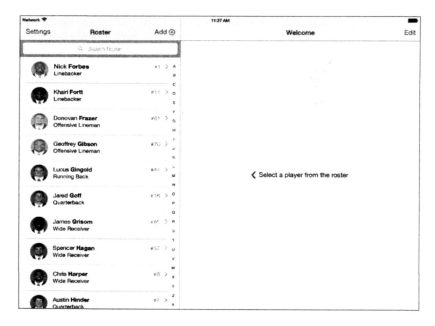

FIGURE 7.9   Hi-Fi Focus prototype screen. (Courtesy of EyeGuide.)

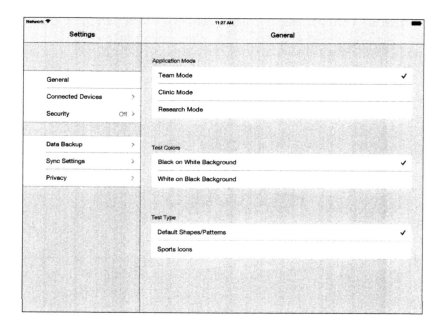

FIGURE 7.10   Hi-Fi Focus prototype screen. (Courtesy of EyeGuide.)

FIGURE 7.11   Hi-Fi Focus prototype screen. (Courtesy of EyeGuide.)

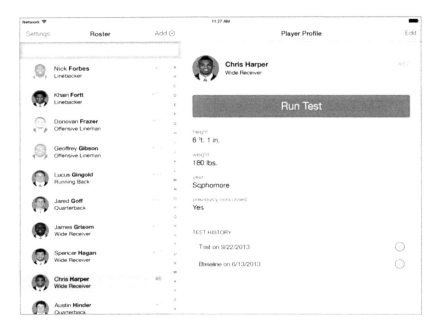

FIGURE 7.12   Hi-Fi Focus prototype screen. (Courtesy of EyeGuide.)

FIGURE 7.13   Hi-Fi Focus prototype screen. (Courtesy of EyeGuide.)

FIGURE 7.14   Hi-Fi Focus prototype screen. (Courtesy of EyeGuide.)

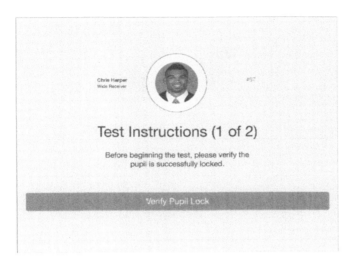

FIGURE 7.15   Hi-Fi Focus prototype screen. (Courtesy of EyeGuide.)

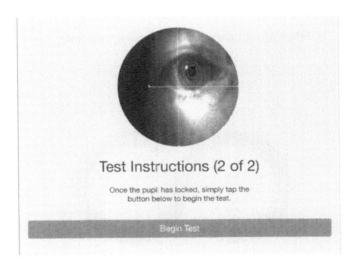

FIGURE 7.16   Hi-Fi Focus prototype screen. (Courtesy of EyeGuide.)

Once developers were convinced that Focus was ready to be tested as a hi-fi, the prototype's interactivity was added using Axure. When athletic trainers were recruited to use this version of Focus, they were told it wasn't a live or real product, but that it was functional and could be used to carry out the tasks they were asked to perform. At this point in development, EyeGuide had learned a great deal from initial user and situation research, all of which was confirmed by a series of tests in which users interacted with Focus prototypes that continued to progress in fidelity at each stage of development and testing.

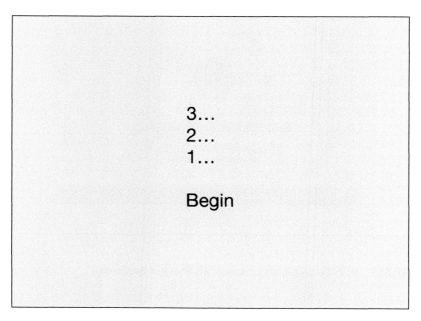

FIGURE 7.17   Hi-Fi Focus prototype screen. (Courtesy of EyeGuide.)

FIGURE 7.18   Hi-Fi Focus prototype screen. (Courtesy of EyeGuide.)

Certain interactive features, such as the results screen experimented within the MVP (Figure 7.3), were not included because they were data-driven based on actual scoring results and thus couldn't be displayed. Instead, red or green circles indicating test failure or success were used as alternatives. This replacement was still used as an opportunity, however, because Focus developers asked

FIGURE 7.19   Hi-Fi Focus prototype screen. (Courtesy of EyeGuide.)

FIGURE 7.20   Hi-Fi Focus prototype screen. (Courtesy of EyeGuide.)

users during testing if they liked the results display and requested alternatives if they did not; they also showed an image of the MVP results prototype to see if users engaged with the hi-fi prototype would agree with the MVP users about how to display results. Good UCD is always seeking to compare multiple sources of feedback and also trusting previous feedback but verifying it as much as possible. Remember UCD Commandment IX: Thou must trust but verify—triangulation is the key.

7.6   INVOLVING USERS IN PROTOTYPING

So far in this chapter, we've looked at the different types of operative images you can create. Now we want to discuss a few different ways you can constructively involve users in the process of designing. We will focus on the following, providing instructions with accompanying examples, for implementing each one:

- Design studio
- Concurrent design
- Collaborative prototype design process (CPDP)

7.6.1   DESIGN STUDIO

Gothelf (2013), in describing how best to implement Lean UX, talks about using a design studio as a technique to encourage collaboration among multiple people that are members of a design team. Certainly, it has that value. As Gothelf writes, the design studio "breaks down organizational silos and creates a forum for your fellow teammates' points of view. By putting designers, developers, subject matter experts, product managers, business analysts, and other competencies together in the same space, and focusing them all on the same challenge, you create an outcome far greater than working in silos allows" (37).

Additionally, the design studio represents a great opportunity for you, as the designer, to include users as part of this diverse, collaborative design exercise. It takes some extra work on your part, and to involve users does mean you more or less know who your users are. But by getting out of the building (GOOB) and working with users as well as other members of your design team to build out a prototype of your product, you will gain a great deal of knowledge that you can use to not only make the product better, but also improve your overall understanding of user needs, wants, and situational constraints.

To implement a design studio, the process is straightforward (Gothelf 2013; Abut 2014), and, as always, you should feel like you have leeway in modifying it to suit your particular design challenges:

1. Define your goals—what do you want to accomplish? What sort of design problem are you trying to solve?
2. Break off as individuals or in small groups to work on addressing design goals; you could also break off with one designer and one user or a small team of designers and user(s).
3. Create something (it's best to put a time limit on each design stage, e.g., one to two hours) and present your findings to the larger team; the larger team will then discuss, offer critique, and set goals for the next iteration.
4. Break off again into individuals/small teams to work. (One twist to this is to move around the users to work with different individuals/teams or even break up small teams and reorganize so designers also move around.)
5. Meet again as a team, present and critique, and then do it all over again until the goal has been achieved.

The obvious, overarching goal is to rapidly iterate a team-driven prototype. The design studio, in principle, is the antithesis of the whiteboard, brainstorming sessions used in some organizations that make participants feel like a lot of work is getting done but in reality, often, to quote Shakespeare, are exercises "…full of sound and fury, signifying nothing" (Shakespeare 2010, 5.5.27–28). Brainstorming on a whiteboard has the tendency to be cognitively chaotic, if not more deconstructive than constructive. Groupthink (when group members agree automatically without much consideration of the problem or task) can take over quickly, or particular voices vie to dominate. Brainstorming can work if moderated carefully, but the design studio offers a useful alternative.

Focus developers adapted the design studio approach by asking users to work with designers to develop multiple potential displays for the results and output of the Focus. The users and designers (one user, one designer per team, three teams total) were first asked to sketch and label what they'd like included on the screen displaying the results. Once they finished, they were asked to repeat the process, but to make their new display unique. After that, they were asked to repeat the process a third and final time. At the conclusion of each stage of the process as well as at the very end, the designers and users discussed as a larger group the multiple sketches created and then derived recommendations for what eventually would become the paper prototype of Focus.

Figures 7.21 through 7.23 show a series of examples from one user/designer team that participated in the Focus design studio. Figure 7.21 shows the initial effort, and Figures 7.22 and 7.23 show the successive iterations built off of it. Note how the first

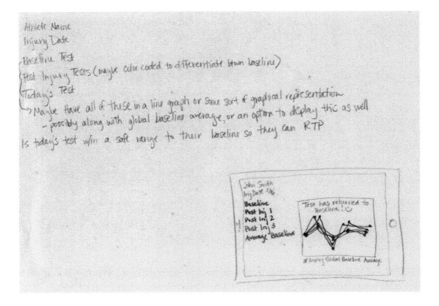

FIGURE 7.21 First iteration of user design during Focus design studio. (Courtesy of EyeGuide.)

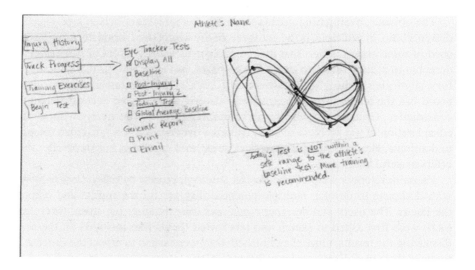

FIGURE 7.22   Second iteration of user design during Focus design studio. (Courtesy of EyeGuide.)

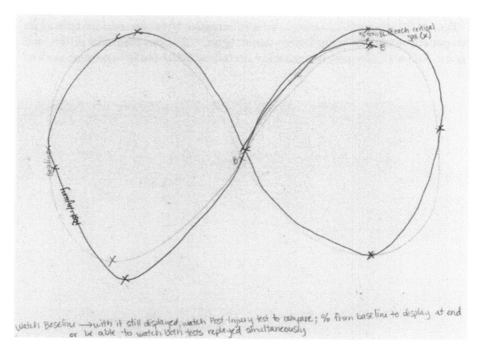

FIGURE 7.23   Third and final iteration of user design during Focus design studio (see how the users' conceptualization of the output matures through successive iterations). (Courtesy of EyeGuide.)

effort is as much text as it is sketching, showing how the user and designer began to formulate ideas about what they thought needed to be on an output screen to report concussion. However, after each discussion with the group, their ideas and sketches became more refined.

7.6.2 CONCURRENT DESIGN

In an effort to iterate rapidly but still involve users, one approach Brian innovated, drawing from Lean UX, is to have users work on new designs as other users work with the existing design. Think about it another way using the police sketch artist as the example. You see someone commit a crime but don't know who it is. You sit down and describe the suspect to the police and, as you provide details and more details are asked of you, the police sketch artist concurrently draws a sketch of the suspect. Once the drawing is complete, you can provide feedback to finalize it, but at the end of the process, a sketch is generated that often ends up on television or the WWW, among other places, asking help from the public to identify the suspect.

Concurrent design works in almost the same way. Brian thought about the approach during paper prototype exercises in a corporate UCD course he was teaching. The corporate course lasted only three days; during that time, the students expected to not only learn UCD, but also incorporate it into the construction of an actual prototype they would use for what would become their new organizational website. The students in the course, all from the same organization, had access to a population of their product's actual users But what they didn't have was a lot of time. If they wanted to iterate a product prototype to be tested and then iterated and tested again while also learning about UCD, all in three days, they had to move quickly.

Often during prototyping, when users are asked to interact, the designer takes notes, returns to the design team, considers possible solutions, and then iterates a new design derived from user feedback. To shorten that process so that the user feedback is integrated faster, Brian's concurrent design approach calls for designers to design as users engage with the product. Go back again to the police sketch artist process for clarification. Because time is of the essence to get a visual representation of the suspect out to the public for help in identification, the sketch artist creates as you describe. Such is the process for concurrent design.

The fidelity of what users work with as you design concurrently can vary. Users can have an initial hand-drawn sketch of the product, the same one you have, and as they work with and comment on it, responding to questions or carrying out tasks, you translate their feedback into design elements on your prototype version. But users may not have a prototype to work with because none exists. In this case, like the police sketch artist example, as the user talks and answers questions, you sketch out a prototype based on user feedback. Should you be okay with stronger user participation, a user also could create the sketch while in conversation with you. Or a more formal prototype may exist, perhaps online, and as a user works with it, you design concurrently. Because it is online, you could design remotely and have multiple designers working at different locations, all designing remotely, concurrently. For any of these scenarios, at the end of the process, you show the user the design created based on concurrent feedback. The user can provide additional commentary and be encouraged also to make modifications. The result is a rapidly developed prototype derived from active user/designer interaction. At the very least, you learn a lot more about your users. But what's also possible is that you can quickly generate worthwhile operative images.

7.6.3 COLLABORATIVE PROTOTYPE DESIGN PROCESS (CPDP)

The last technique for managing a design process with multiple designers and users is the collaborative prototype design process, or CPDP. Created by a large group of graduate students in one of Brian's graduate usability courses, the CPDP was refined after this group's initial experimentation, becoming a published approach (Andrews et al. 2012) that is now used by many designers (Corneto 2012).

The CPDP, like the design studio, is a technique for managing the collaboration of multiple designers, along with other stakeholders, in the development of a product. It initially solved a practical problem for the inventors of the technique. There were 12 students in the graduate course, and all were responsible for contributing to the development of a new website. It wasn't effective to have all of them to work together to build out one prototype, then test just that one prototype and revise it. For one, they couldn't learn hands-on design principles if they were forced to share the limited resources. Some students would inevitably be left out of trying something out because there weren't enough computers or other technologies to go around. This was also a practical project for a real client, and having 12 developers only work on one prototype or conduct one test didn't maximize the potential of having them split up and create comparative prototypes as well as carry out comparative testing with different sets of users.

The CPDP approach, which they came up with to formalize the practical solution they experimented with to be effective during the rapid development of the new website, has a few repeatable, straightforward steps:

- Identify the user group.
- Develop and test three independent paper prototypes for the same website.
- Evaluate and interpret data.
- Create three wireframes of a website.
- Test those wireframes for usability with representative samples of users.
- Gather and analyze data per team and collaboratively across teams.
- Merge designs to create one final, user-driven prototype of the website. (Andrews et al. 2012, 125)

What makes the CPDP different from the design studio is that the smaller teams initially create paper prototypes separate from the influence of other teams' creations. As a means of stifling any potential for groupthink, the smaller teams do not consult with each other about the designs they create until they have developed prototypes and then tested them with users. All small teams work with users recruited from the same user profile, but they don't share design decisions or test outcomes until the completion of the paper prototype phase.

In addition to design independence among the groups, initially, the CPDP also differs from the design studio in its formality. Three distinct phases—paper prototype, wireframe, and a final hi-fi prototype—must be carried out as part of the process. The design studio is less structured, allowing designers to choose to carry out as many prototyping sessions as necessary. Finally, the CPDP does not involve users as part of the design team. Users strictly provide feedback to design teams by using the product prototypes as part of usability tests at the end of each design phase.

Ultimately, you may choose the CPDP if you have a larger group. Andrews et al. (2012) note that "the process requires more people than an individual designer or a small design team would normally require" (139). In fact, EyeGuide, as a startup, has a small design team and so lacks the luxury of being able to spread its people into three semi-independent teams. However, for organizations that have more mature operations, have more staff on hand, and want to have a more formalized process in place for carrying out collaborative UCD, the CPDP is a great technique. At least six prototypes are generated quite rapidly, followed by a seventh hi-fi prototype, which is derived from user testing and subsequent team discussion of the previous prototypes. Such testing can gather quantitative and qualitative data, and this mix may be more appealing than the rougher, more heavily qualitative data the design studio often generates.

The fact is that you have a lot of options available to you to generate an operative image, and you should plan for what works best for you given your team size and your time constraints as well as other factors, such as knowledge of users, availability of users, materials, etc. Ultimately, you must build out something for users to use, learn from that, and improve what you built to move forward in the process of iterating a UCD product.

And we now also move forward. You've learned how to build out an operative image. It's time for the last part of the RABBIT: let's learn how to test with users to get great feedback.

TAKEAWAYS

1. An operative image or prototype that has the look and feel of your eventual finished product allows you to test its effectiveness with its intended users.
2. You can build out an operative image even in the midst of researching users or carrying out other types of knowledge gathering if it aids in determining user needs and wants. (The RABBIT process isn't necessarily linear.)
3. The minimum viable product (MVP) approach is a constructive way to create a prototype as fast as possible, have users use it, then make changes to it and test it again. It is also a fast way to test a hypothesis about your product and helps answer the question, "Do users need and want it?"
4. Paper prototyping allows you to involve users to interact with a simulation of the product early on in the process. Paper prototyping encourages modification early on and saves on costs of expensive changes later.
5. A high-fidelity test allows users to use most product functions, but it doesn't necessarily tell you why users failed to complete certain functions.

Conversely, a low-fidelity test generates more feedback about users' mental models and how they might expect a product to function.

6. Whether you conduct low-fidelity or high-fidelity testing, you need to have a purpose for the test. Why are you asking users to interact with this prototype?

7. It's important to create a scenario, a story that requires the users in question to carry out tasks. The scenario and tasks should be representative of the actual use environment.

8. A concurrent design approach calls for designers to design as users simultaneously engage with the product. The result is a rapidly developed prototype derived from active user/designer interaction.

9. Collaborative prototype design testing can gather quantitative and qualitative data; this mix may be more appealing than the rougher, more heavily qualitative data that the design studio often generates.

DISCUSSION QUESTIONS

1. Think of an MVP for a new banking app you're developing. What functions would it have? What functions would be left off for future iterations or until user feedback determined whether they're necessary?

2. Using the banking app example, which features and functionality might a low-, medium-, and high-fidelity prototype have?

3. Using the banking app example, what scenario or tasks might you create for users? What information would you use to develop these scenarios or tasks?

4. If you had only the time, money, and resources to complete low-, medium-, or high-fidelity prototype testing, which type of test would you recommend?

REFERENCES

Abut, A. "How to Run a Design Studio in 90 Minutes or Less." *Zapier.* February 4, 2014, accessed February 25, 2016. https://zapier.com/blog/run-a-design-studio/.

Andrews, C., D. Burleson, K. Dunks, K. Elmore, C. S. Lambert, B. Oppegaard, E. E. Pohland, D. Saad, J. S. Scharer, R. L. Wery, M. Wesley, and G. Zobel. "A New Method in User-Centered Design: Collaborative Prototype Design Process (CPDP). *Journal of Technical Writing and Communication* 42, no. 2 (2012): 123–142.

Armstrong, M., and M. Olson. "EyeGuide Test Plan." Unpublished corporate report, 2015.

Corneto, T. "Review of the Collaborative Prototype Design Process (CPDP)." *Blendesigner.* October 8, 2012, accessed February 25, 2016. http://blendesigner.com/2012/10/28/review-of-the-collaborative-prototype-design-process-cpdp/.

Garrett, J. *The Elements of User Experience: User-Centered Design for the Web and Beyond.* Berkeley: New Riders Publishing, 2011.

Gothelf, J. *Lean UX: Applying Lean Principles to Improve User Experience.* Sebastopol: O'Reilly Media, 2013. Kindle Edition.

Kelly, R. "Minimum Viable Product: What It Is and How to Build One." *Rob Kelly.* April 13, 2011, accessed February 25, 2016. http://robdkelly.com/blog/entrepreneurship/minimum-viable-product-mvp/.

Lowgren, J., and E. Stolterman. *Thoughtful Interaction Design.* Cambridge: MIT Press, 2004.

Medero, S. "Paper Prototyping." *A List Apart.* January 23, 2007, accessed February 25, 2016. http://alistapart.com/article/paperprototyping.

Rettig, M. "Prototyping for Tiny Fingers." *Communications of the Association for Computing Machinery* 37, no. 4 (1994): 21–27.

Ries, E. *The Lean Startup.* New York: Penguin Books, 2011.

Shakespeare, W. *Macbeth.* Boston, English Play Press, 2010. (Original work published 1699).

Smith, D. "One of the NFL's Top Quarterbacks Trains with Virtual Reality." *Tech Insider.* November 18, 2015, accessed February 25, 2016. http://www.techinsider.io/nfl-virtual -reality-2015-11.

Snyder, C. *Paper Prototyping: The Fast and Easy Way to Design and Refine User Interfaces.* San Francisco: Morgan Kaufmann, 2003.

———. "Paper Prototyping." *Human-Computer Interaction.* November 1, 2001, accessed February 26, 2016. http://www.cs.uu.nl/docs/vakken/hci/index.php?id=3&id1=0&id2=7.

Still, B. "Making Our User Experience More Lean." *Intercom* 61, no. 7 (2014): 27–28.

Still B., and J. Morris. 2010. "The Blank-Page Technique: Reinvigorating Paper Prototyping in Usability Testing." *IEEE Transactions on Professional Communication* 53, no. 2 (2010): 144–157.

8 Test the Design

Once you've created a prototype, you need to test it to gather real feedback from real users. You'll remember our emphasis on testing in Chapter 2 and UCD Commandment IX: Trust but verify. Usability testing helps you determine if users can complete tasks using your prototype(s) and verify whether your product is user friendly.

When conducting usability testing, you need to consider the goals of the design team, the goals of the users, and the product's performance toward these goals. By using an iterative design approach, you integrate the knowledge you acquire from frequent testing to improve the design as it moves from prototype, to beta design, to version 1.0, and beyond. As UCD Commandment X states, discovery never ends. Testing the design with users, getting feedback, and then improving both the design and users' knowledge is a major driver in the ongoing discovery process.

To learn more about usability testing as a central part of fundamentally sound UCD, we discuss the following in this chapter: defining and measuring with usability testing, choosing a usability test, creating a test plan, conducting your test, and analyzing your data.

8.1 DEFINING AND MEASURING WITH USABILITY TESTING

Usability testing is a research method that analyzes how well the product being tested is used by its intended users. The mantra of usability testing is that you test representative users performing representative tasks. Like user research (discussed in Chapter 4), we want to observe the *see-say-do triangle*. Use the *see-say-do triangle* in usability testing just as you would for user research: Observe what users do, listen to what users say, and measure what users do. This means you balance observation, self-reporting, and quantitative data. Further, you use the guiding principles of MEELS to help us focus our data collection. When you perform usability testing, you are looking to understand five usability quality components: *M*emorability, *E*fficiency, *E*rrors, *L*earnability, and *S*atisfaction (MEELS). These components aim to understand the following questions:

- Memorability: How well do users remember how to complete tasks so they may use this information to complete new tasks? This may include knowing which menus to use to find information on websites or software or even how to switch between apps on a smart phone.
- Efficiency: How long does it take users to perform tasks? Is this a reasonable amount of time? What prevents users from performing tasks more efficiently?
- Errors: What errors are made while performing tasks? How severe are these errors? Can users recover easily from errors, or do the errors lead to task failure?

- Learnability: How well do users learn the system? Do they perform tasks more efficiently in later tasks (once they've had a chance to learn the system)?
- Satisfaction: Do users like the system? What about the system do or don't users like? (Nielsen 2012a).

Consider MEELS a balancing act: You want your product to achieve all the elements, not just one. Sometimes, user satisfaction is confused with actual usability, but preference is not a usability measurement. Your users may want features in your design that are not necessary to use the product—this is a preference, not a need. Using MEELS will help you determine the elements of your test design and the usability of your product (we discuss this more in a bit).

Further, keep in mind that you are always testing the system. In other words, it may be tempting to blame errors and lack of efficiency on users. However, you are testing the product for a reason. You want errors to be revealed. You want lack of efficiency and the reasons for this lack of efficiency to reveal themselves. These "failures" are opportunities. If you can isolate problems, you can incorporate what you are learning into the new stages of your design; thus, your design continues to mature and improve.

8.2   CHOOSING A USABILITY TEST

There are two main types of usability testing: summative and formative. Summative usability is more experimental, uses larger sample sizes, and is concerned with achieving statistical significance. Summative testing is used to conduct comparison testing (often referred to as A/B testing) or baseline or benchmark testing. Baseline or benchmark testing is when the designer aims to determine an average performance measurement for a specific product to determine what a "normal" range would be. This type of testing requires larger sample sizes for each individual test and the running of statistical analyses (see Sauro and Lewis 2012 for more information on determining sample sizes and statistical tests). Formative testing uses smaller sample sizes but tests more frequently during the design of a product. For UCD, A/B testing and formative testing are the two most useful tests for comparing and testing products during your design process.

8.2.1   A/B TESTING

A/B testing compares the usability of two competitive products, two prototypes of a product, or even an earlier versus a later version of a product. Is one product or version used more successfully than another? Why? What does this say about how you will begin or continue designing your product? Although this testing does require a larger sample size than formative testing, it can be useful in understanding how users respond to different design options. Nielsen (2012b) defines A/B testing as a low-cost test to understand how one interface performs against another.

For instance, if you want to see whether a new web design is likely to solve navigation issues, test users with navigation tasks for design A and design B. You need at least 20 users for A/B testing, which is considered a quantitative study (Nielsen 2012c). After testing users with both designs, you need to run a statistical analysis to determine which one shows significant change (usually a 90% confidence level is enough to determine significant difference). Whichever design has the best statistical performance is the design you should stick with (also see Macefield 2009).

There are limitations to an A/B test. For one, it is not meant to test the full range of usability components because it makes determinations on quantitative data; this means you would not see user behavior patterns when you analyze data. A/B data has its place; it can be highly effective for understanding efficiency patterns and navigation trends, but your use of A/B testing should be balanced with additional usability testing that incorporates both quantitative and qualitative data.

8.2.2 FORMATIVE TESTING

Formative testing, also known as iterative testing, is perhaps the most useful type of testing to conduct during your design process. Formative testing emphasizes testing multiple times during the design process, using small numbers of representative users to test for each iteration. Because of the small sample size, people refer to this method as discount or guerilla usability. However, the assumption that this method is less valuable than larger participant testing is misguided. We have already discussed the need to balance methods with resources, and you will find that this is true of usability testing as well. By testing small numbers of users multiple times while designing your product, you get data about usability problems that can be fixed during the design process. If you waited until the end of the design process, when you are ready to send your product out for use, you run the risk of major design flaws and usability problems that may need to be fixed or cannot be fixed because the design is too close to production for major changes. Thus, as emphasized throughout this book, spreading out your resources and testing frequently during design is the best way to circumvent major usability problems when you are ready to release your product for use.

Iterative testing pairs best with UCD because it acknowledges that users are integral to the design process: Without user input, how can you be sure your design is appropriate for representative users? Thus, the goal of iterative testing is to test multiple times during design.

How often? As often as you can, given your schedule and goals. We suggest, as Figure 8.1 shows, that you test a paper prototype of your product, a wireframe, and then a hi-fi or beta prototype. If you can, test also the final 1.0 version of your product. In fact, continue testing after that, using the data you collect to continue to improve the product or gain knowledge to make other products (Morgan and Borns 2004; Bergstrom et al. 2011).

For formative testing, Nielsen (1993) recommends using at least five users for each iteration of testing. The major argument against testing this small of a sample is that not all errors will be caught by these five users (Caulton 2001; Faulkner 2003; Spool and Schroeder 2001). However, if testing while designing, you will perform more than one test. Nielsen (1993) explains,

It is likely that additional usability problems appear in repeated tests after the most blatant problems have been corrected. There is no need to test initial designs comprehensively since they will be changed anyway. The user interface should be changed and retested as soon as the usability problem has been detected and understood, so that remaining problems that have been masked by the initial glaring problems can be found.

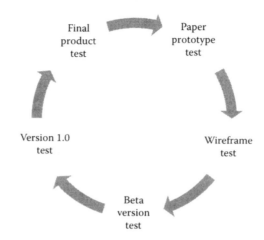

FIGURE 8.1   Iterative testing cycle.

Additionally, you likely won't have the time nor will it be your goal to conduct a large-scale, statistical, randomized sample of hundreds or thousands of users to find problems that may or may not affect most people in the world. You want to make the best product possible with all things considered, such as time, money, client goals, situation, and user needs and wants. You test users to integrate them into the UCD *design process*. In all likelihood, by the time you have conducted tests for a paper prototype, a wireframe prototype, a beta version of your product, and a final version, you will have tested dozens of users. And, if you continue testing beyond the final design, which you should, those are additional user tests. If you have the resources (e.g., time and money) to test more users for each iteration, go ahead. But you can gather a lot of great data from a small number of users as long as they are representative of your users and you involve them throughout.

8.3   CREATING A TEST PLAN

You know you need to test your design, but where do you start? Testing isn't as easy as bringing in users and watching them work. It takes careful planning to ensure you are testing the right users, the right tasks, and under the right conditions. The time you spend planning not only leads to smoother testing conditions, but also is evidence to your team and clients that the test you performed accurately represents the goals of the project and follows appropriate, consistent protocols. For this reason, we discuss planning the test by breaking down how to write a test plan for your test.

Developing your test plan is much like answering journalism's who, what, when, where, why, and how questions although not necessarily in that order. Essentially, you want to clearly indicate the goals of your test (what), your user profile for the product you are testing (who), the reasons and context for testing (why), the time frame for conducting the test (when), the location and environmental conditions of your test (where), and the steps for carrying out the test (how). The following sections will break down each of these elements of the test plan (see Box 8.1 for a checklist),

BOX 8.1   TEST PLAN CHECKLIST (COURTESY OF EYEGUIDE)

TEST PLAN CHECKLIST

Use this as a checklist. If you've answered every question, your test plan will more than likely be ready to go.

- ❑ **Purpose**
  - Why are you doing the test?
  - What are your team goals, the goals of the client for this stage of testing?
- ❑ **User Profiles**
  - What are the different user groups?
  - How do you define them? What are their most significant representative characteristics?
  - Which user group(s) will be tested during this phase?
- ❑ **Methodology**
  - How will you test?
  - What methods will you use to gather data from the users before, during, and after testing? Remember the triangle
- ❑ **Scenarios/Task List**
  - What scenarios have you created that the users will be asked to operate in during testing?
  - What tasks will you ask them to perform?
    - Typically five to seven closed ended or a mix of open and closed
    - Studies show users degrade cognitively after 30 minutes or so of testing
    - Don't ask for a task unless it is something a user would do and it tells you something about the user experience that matches your overall test goals
- ❑ **Evaluation Methods**
  - Will you just collect the data and then analyze from the bottom up (looking for trends or affinity in the data) after the fact, or will you look for certain kinds of data while the testing is going on?
  - What kinds of data will you collect?
  - Are you trying to confirm where your product stands against others? Are you validating changes you put in? Do you have hypotheses you want to test?
- ❑ **Test Environment**
  - Where will you test?
  - How will you make the environment resemble a realistic setting?
  - How will you record or collect data?
- ❑ **Test Schedule**
  - When will you test?

- How long do you anticipate testing to occur?
- How many users do you intend to recruit, and how will you recruit them?
❑ **Team Members, Roles**
  - Who is considered to be a member of the test team?
  - What are the roles on the test team?
❑ **Plans for Reporting**
  - What will you do with your results?
  - How do you intend to report them?
❑ **Appendices**
  - Script for Facilitation
  - Pre- and Post-test Survey Materials
  - Observation Logs (or marker tables if using testing software, such as Morae)

describe each component, and provide an example from the EyeGuide Focus project's paper prototype test plan (prepared by Megan Olson and Miriam Armstrong).

8.3.1 TEST GOALS

Your test goals are the desired outcomes from the test—not the desired results, but what you want to learn about your product. These goals can include learning about design elements in a current prototype, observing how users would perform tasks using a prototype, testing a beta version of the product to detect errors, or finding out which of the A/B product versions users use better. These goals can be written as statements or as research questions. In Box 8.2, the designers indicate two design goals for the Focus prototype test: to confirm assumptions made during discovery research and to get feedback on initial prototype designs.

BOX 8.2   TEST GOALS FOR EYEGUIDE FOCUS PROTOTYPE TEST (COURTESY OF EYEGUIDE)

The goals of this paper prototype test are the following:

1. To determine whether the user requirements elicited during the discovery phase are consistent and complete across a wider group of users.
2. To procure feedback on an initial design in order to evaluate the design and inform successive iterations. Specifically,
   a. What information do trainers use to make decisions about whether the athlete should continue rehabilitating or see a physician to resume play?
   b. How do the trainers interact with the results display?

8.3.2 CONTEXT FOR TESTING

This is essentially a discussion of preliminary research, or discovery, you have conducted to understand your product. This information should include user research, product research, design decisions that have been made based on this research, and findings from previous tests. This provides your team and client context about the progress of the project and how the upcoming test will move both your and their understanding of the design process forward. Box 8.3, for instance, discusses the research that went into understanding the user's current process for detecting concussions. Further, the designers discuss the design methods used after they collected interview data to design a prototype with their interviewees' participation. These design models were then used to define a user profile and tasks for testing.

**BOX 8.3   CONTEXT FOR EYEGUIDE FOCUS
PROTOTYPE TEST (COURTESY OF EYEGUIDE)**

Before beginning initial testing and prototyping, we met first jointly and later individually with our initial user contacts, N & K, both high school athletic trainers. During our first meeting, we investigated the current system for concussion detection and monitoring. We learned about the current processes for identifying the concussion by asking N & K to talk through the process of treating someone who may have sustained an injury (modified critical incident analysis). Additional probes included questions about information needed to make a decision, artifacts used during preliminary diagnostics, other individuals involved in the process, and current monitoring procedures. We also asked about levels of experience, training, education background, and the safety culture of the school and athletic system. Thus, our initial interview served to develop our user profile.

Following the interview, we continued with a site visit to observe N collecting baseline information on athletes. By observing her current method of data collection and actively asking questions (for example, "What information are you considering to decide that this test is 'okay'?"), we began to understand the context in which baseline information would be tested, the user's mental model of how the test worked and what was qualitatively "good," and the existing interface with which our results screen would need to connect.

Our second meetings were with each trainer individually and were intended to help us select which display options we would use to begin prototyping. We adapted a Lean UX (Gothelf 2013) strategy called design studio by asking the users to develop multiple potential displays for the results and output of the eye-tracking data collection. The users were first asked to sketch and/or label what they'd like included on the screen displaying the results. (We acknowledge the recommendation to avoid asking users to be designers, but decided this was a valuable opportunity to generate display requirements.) Once they finished, we asked them to repeat the process, but to make their new display unique. After that, we asked them to repeat the process a third and final time.

At the conclusion of the three iterations, together with the user and all three designs in front of them, we asked the user to describe what they included in their design and why, the strengths of each design, how they'd use the displayed information to make a decision, and, after that discussion, what changes they'd make to any of their designs.

Based on these designs, we came to a better understanding of the following:

- What the users were looking for to make their decisions (e.g., symptoms, comparison to individual baseline)
- What they valued in a display (e.g., quantitative numbers for a comparison, visual representation of the data, and an evaluation of the student's status of deviation from normal performance)
- Their mental model of the type of data being collected (they weren't looking for "jerkiness," but rather the shape of the "figure 8" in the video)

By the conclusion of these meetings with our user contacts, in collaboration with information from EyeGuide concerning system capabilities and constraints, we established a starting point for our initial prototyping tasks and a better understanding of who our users might be and in what environment they may be using the system.

8.3.3 USER PROFILE

You should have copious user data by the time you begin testing phases. Chapter 4 provides multiple user research methods to design personas and determine a user profile. For the test plan, you should isolate the type of representative users needed for your test. You may take major characteristics from the personas you've developed to include in your profile so you can target specific user groups (Box 8.4).

BOX 8.4   USER PROFILE FOR EYEGUIDE FOCUS PROTOTYPE TEST (COURTESY OF EYEGUIDE)

According to information provided by EyeGuide, the target user population for this product will be athletic trainers at the middle school and high school levels. Based on discussions with EyeGuide and initial conversations with users, others who could potentially interact with this system would be student athletic trainers and coaches. Student athletes, parents, and physicians may also have the opportunity to view the output of the system at different times, but for the upcoming iterations, high school and middle school athletic trainers will be the target population.

8.3.4 User Recruiting

Recruiting participants is one of the most difficult parts of testing. Therefore, you want to outline answers to the following questions:

1. How many users will you test?
2. Which user groups (from your user profile) will you test?
3. How will you find participants for testing?
4. What incentive is necessary to persuade participants to test?

How many users you test should be determined by the type of test you conduct. If you are conducting a formative, iterative test (such as the Focus prototype test example used here), at least five users should be used. You can break down these users based on the personas you have created in your user research. One user may be a 25- to 30-year-old female athletic trainer at a midsized high school with zero to five years of experience; another may be a 35- to 40-year-old male athletic director at a large middle school with 10+ years of experience. The key is to find participants that represent the users you have described in your user profile. However, if you are conducting A/B testing, at least 20 users will be needed to determine statistical significance. If you are conducting a larger scale test, you also need to determine how you will recruit users to represent each user group. You may choose to use more detailed stratified sampling in which you not only break down user characteristics, but also divide the number of participants per characteristic profile by the representative users with those characteristics (Sauro 2012). The most important points to remember when determining users and sample size are (1) participants need to be representative users, and (2) the sample size needs to be appropriate for the type of test you are running.

You can begin recruiting by using resources already available to you. First, it helps to know contacts in the user community that can help you find representative users. These contacts may be people who you have worked with in other phases of user research, such as site visits, focus groups, or interviews. Consider your own network of professional contacts to help you find representative users. There is also the option of working with services such as usertesting.com that can connect you to remote users for testing. Although not free, these services are good options if you are looking to expand your participant pool beyond your physical location (Box 8.5).

People are busy, and asking them to take time to test may require additional incentive. Some incentives commonly used are paying participants for their time, offering gift cards for participation, entering participants in a random drawing for a prize (such as a larger value gift card or iPad Mini), or even offering a free trial for the software you are testing. Of course, you have to balance your incentive with your financial resources, but having an incentive can greatly improve your recruiting efforts.

> ## BOX 8.5   USER RECRUITING PLAN FOR EYEGUIDE FOCUS PROTOTYPE TEST (COURTESY OF EYEGUIDE)
>
> User recruiting will be performed by reaching out to athletic trainer contacts provided by N and located within reasonable driving distance of Lubbock. We will ask those athletic trainers to participate in our iterative rounds of testing and also request additional contact information for other trainers in middle and high schools to increase and diversify the users available for testing. Screening questions will be included in this stage of contact to avoid arranging testing times with ineligible users.
>
> We intend to recruit five users for this stage of testing.

8.3.5   TEST PROCEDURE

8.3.5.1   Description of Design for Testing

What are you testing? Here, you'll describe the product being tested. Is it a prototype? What does the prototype look like now? In this case, images can clarify the exact product being tested (and can help you keep an accurate record of which test plan goes with which design).

8.3.5.2   Test Environment

The test environment refers to the conditions under which testing occurs. This should include date ranges and times (i.e., schedule for testing), location for testing, location setup, equipment needs and setup, etc. Indicating a test schedule should clearly set a block of time for you and your team to focus on testing. A conservative amount of time for a simple test of about five tasks is about an hour per test. Participants probably won't need this amount of time; however, an hour provides you time to set up, test, save recordings and observation notes, and prepare for the next participant. If your test is more complex, you may want to consider lengthening the amount of time scheduled for each participant.

You also need to decide on the best location for testing. Usability testing can be conducted in various locations: in a lab, in the field, in an off-site dedicated space, or even remotely. What location best represents the users' environment and allows for the best testing conditions? In the case of paper prototyping, you will need space for users to work with paper. This might be a lab or conference room with tables where you can move papers around. If you are testing a large number of participants using software, then a lab space where it is easy to control conditions may be the best option. Finally, if you want to see how the design works in the user's environment, you need to test in the field or remotely.

When you know where the test will be located, consider the setup for each test: what equipment and materials are needed, how equipment and furniture should be arranged, and where team members will be in relation to users and equipment. Your list of materials should include all items that the participant and team will engage with, such as the following:

- Consent forms
- Pretest survey (if not loaded on computer)
- Computers for testing
- Computers for observation
- Computer components (keyboard, mice, etc.)
- Mobile devices
- Software or apps to be loaded on computer/mobile device for testing
- Software or apps to be loaded on computer/mobile device for recording and observation
- Paper prototype materials (paper, writing utensils, sticky notes, tape, etc.)
- Table
- Chairs

This list is by no means exhaustive, and your own list will be much more specific to your testing goals.

8.3.5.3 Team Roles

Although testing can be accomplished with just one experimenter, having a team of researchers at your disposal can make for a smoother process. Basic usability teams consist of a facilitator and one or two observers. The facilitator has direct contact with the participant. The facilitator is in charge of communicating to and collecting information from the participant such as introductions, preliminary data collection, answering questions, reading the test script, and asking posttest interview or retrospective recall questions. Observers are responsible for taking notes on user actions, behaviors, and verbalized content. Observers can be in the same room as the facilitator and participant, or they may observe from another room. The location of the observer depends on the environment of the testing location.

In Box 8.6, you see the usability team created another type of role: the computer. In paper prototype testing of digital designs, participants are asked to manipulate

---

BOX 8.6   TEAM ROLES FOR EYEGUIDE FOCUS PROTOTYPE TEST (COURTESY OF EYEGUIDE)

- Each of the experimenters will alternate in filling the role of moderator or computer.
- The computer will be responsible for manipulating the paper prototype in response to the user's actions. This might include adding "popup" menus, changing windows/screens, "scrolling," etc.
- The moderator will be responsible for directly interacting with the participant, following the script, encouraging the user to continue verbalizing throughout the test, asking questions as part of active intervention, and handling user questions and comments.
- Both the computer and moderator will be taking notes, but the moderator will hold primary responsibility for note-taking throughout the test.

paper "screens." Therefore, one team member needs to act as the computer or digital device, advancing pages if users click on links or submit buttons.

Team roles can be fixed or fluctuate throughout testing. For the test plan, it is important to note what team members are needed and the roles and responsibilities for each.

8.3.5.4 Test Tasks

Your test's tasks, like its users, should be representative of tasks that users would perform using your design. There are two options for creating tasks. First, you can create tasks and provide data for completing those tasks. The benefit of predefined tasks is that you can control the test to better compare data among participants during analysis. If you choose to create tasks for your users, which is most common in usability tests, you should formulate tasks based on your user research: What will they be using the product to do? Each task should have clear starting and ending points (refer to Chapter 4's task analysis discussion). Box 8.7 shows examples of tasks that represent what users would use the EyeGuide Focus device for as well as tasks that have a clear beginning and end for each.

The second option is to allow participants to create their own tasks and use their own data to complete these tasks. For instance, if you are designing a new budgeting app, users could interact with the app using their own information and perform tasks without prompting from the facilitator. Genov, Keavney, and Zazelenchuk (2009) advocate using real user data in usability testing rather than giving data to users to use during the test to increase user investment, reduce cognitive load, and ensure realistic data is used. Further, during early design stages, Hawley (2012) calls for user-derived tasks that provide feedback on not just how usable the product is or could be, but also what users would use the design for: "The goal is to understand whether the proposed design aligns with user needs and expectations. Even if the design artifact or prototype does not support exactly what they want to do, I can learn from their intentions and explorations" (par. 16).

Whether you choose to create tasks and data or have user-directed tasks and data, the tasks should be appropriate for investigating your test goals. Keep in mind that,

---

**BOX 8.7   TASKS FOR EYEGUIDE FOCUS PAPER
PROTOTYPE TEST (COURTESY OF EYEGUIDE)**

Here we summarize each of the tasks the user is asked to perform with the prototype. Note that Tasks 1 through 4 will refer to one hypothetical athlete (athlete A) and Task 5 will refer to a second hypothetical athlete (athlete B).

- Task 1: Indicate a baseline for athlete A.
- Task 2: Select tests to compare baseline to athlete's recent input.
- Task 3: Decide if athlete A should continue rehab or see a physician.
- Task 4: Rename a test.
- Task 5: Decide if athlete B should continue rehab or see a physician.

as you perform additional tests throughout your iterative process, goals for each test will change. Thus, the tasks you develop for the test will also change.

8.3.6 Data Collection Methods

How will data be collected during the test? We are looking for data to answer questions or provide insight for our research goals. Further, we need to select data that fits into our *see-say-do triangle*. Remember that "see" refers to observing users, "say" refers to listening to user feedback, and "do" refers to collecting data on user performance. Whether you conduct a prototype test or test a near-completed product, you can use some or all of the following data to understand users' experiences.

8.3.6.1 See

The "see" in the *see-say-do triangle* means that the designers need to observe what users are doing while interacting with the design. Observation data can include observing user navigation with the design and user behavior (e.g., emotional responses, body language). You can also use eye tracking to literally "see" what users do. Eye-tracking data can show where users are fixating on a screen (or design) as well as indicate users' "areas of interest" when using the product. Specifically, eye tracking can provide data, such as gaze plots (or scan plots), area clusters, and heat maps. Gaze plots show the eye's navigation pathway for finding information by numbering the area of fixations sequentially. Area clusters calculate the areas of design that users focus on the most. Heat maps aggregate users' eye tracking to show their areas of fixation (for more information on eye tracking, see Cooke 2005; Duchowski 2007). Ultimately, you want to choose observation methods that relate to your test goals. If testing a prototype, eye tracking probably would not be very useful as you are trying to see how users work with or help to construct a new design. However, if testing a completed software design that seems to have usability problems, eye tracking may help pinpoint areas of the design that are being fixated on or ignored. This, in turn, can help reveal why usability problems are occurring.

8.3.6.2 Say

To find "say" data, you are looking for user feedback about the design. Data collection tools for feedback include think-aloud protocol, active intervention, retrospective cued recall, posttest interviews, and satisfaction surveys. As discussed in Chapter 4, think-aloud protocol (TAP) asks users to talk about their thoughts and decision-making processes while completing tasks. In usability testing, this can be difficult for users. Because they can be nervous or busy focusing on completing tasks, they will need to be prompted to speak if they are being quiet. The facilitator should refrain from answering questions about the task; rather, he or she should prompt users to speak if they remain silent for five seconds as well as use vague, neutral language, such as "uh huh," with them during think-aloud protocol (Ramey et al. 2006).

If TAP is too difficult for users during the test (such as if you are working with children or if testing creates too much of a cognitive load for users), active intervention is a good alternative. With active intervention, you create a series of questions to be asked during the test to collect data about how users completed the task or used a certain tool within the design. Active intervention questions are usually asked at the end of the task or after a specific tool is used. For instance, after a task is completed, you might ask a user, "What would you do next?" Active intervention questions should be placed in your test script, and the questions, and placement of questions, should be consistent for all tests.

Retrospective cued recall is another option for gaining user feedback. In cases in which both TAP and active intervention would be difficult to conduct, retrospective cued recall is a good alternative. Tools such as eye-tracking equipment or virtual reality equipment could cause a cognitive overload if you are also asking users to think aloud or interrupting them while they work. Retrospective cued recall waits for users to finish the test before asking for feedback. At the end of the test, you would replay parts of the test (a specific task for instance) or verbally remind users of actions they took at specific points in the test. You can then ask users to talk about what they were thinking, what information they were looking for, or what prevented them from taking another action. This method is effective in capturing users' thought processes when in-test verbal feedback is not feasible (Russell and Oren 2009).

A last option for obtaining user feedback is posttest interviews or questionnaires. Interviews can be used to gauge user satisfaction after the test. For instance, you may ask users what they liked most about using the system or how easy the system was to use. You can ask them to expand on their answers to get more specific information about their experience using the design. Another popular method for collecting posttest satisfaction information is to use a usability satisfaction questionnaire. One of the most popular surveys is the system usability scale (SUS), a Likert scale to measure user satisfaction. Another questionnaire is the computer system usability questionnaire (CSUQ), which is longer than the SUS, but each statement about the system's usability is more specific (Barnum 2011). Further, whereas the SUS questionnaire is completed immediately after a usability test, CSUQ can be used "across different user groups and research settings" (Sauro and Lewis 2012, 226).

8.3.6.3 Do

Finally, you need to collect "do" or performance data. Performance data are quantitative and can be counted. For instance, time on task, mouse clicks (or number of inputs), error rates, and dwell time are all performance measures. Time on task is exactly what it sounds like: You note the amount of time users spend completing each task. Mouse clicks, also referred to as number of inputs if users are not using a mouse, count the number of times users make a selection per task (such as clicking on a link, clicking a submit button, changing formatting options, etc.). Error rates count the number of errors and severity (which is discussed later) that occur during a task. Finally, dwell time is the amount of time between inputs, which shows how long it takes users to make decisions before taking another action.

The methods above represent some of the most common usability data collected during testing. You should not attempt to use all of these methods; it would overwhelm users and be difficult for the facilitator and observers to keep up with. However, you should choose data that helps answer your research questions or provide insight into your user goals. Further, as mentioned above, you need to be careful of cognitive overload, which occurs when users are asked to do too much or are affected by multiple outside influences tapping their cognitive resources, all influencing their ability to perform tasks. For example, if eye tracking was used while users performed TAP or answered active intervention questions during a test, they would certainly encounter cognitive overload.

One way to decide which methods to choose is to examine what elements are most useful to measure MEELS for your test. You can collect specific types of information based on MEELS. The following is an example of how you can break down data collection using MEELS:

- Memorability: Navigation (e.g., mouse clicks), error, and verbal data. Do users' performances seem to improve in later tasks? Does verbal data indicate that users remembered (or didn't remember) how to perform similar tasks or find information in later tasks?
- Errors: Error and verbal data. What errors do users make, and what thought processes or reasoning guides the different actions that resulted in error?
- Efficiency: Time-on-task data. How long does each task take? Does efficiency improve as tasks are completed?
- Learnability: Time-on-task, navigation, error, and observation data. Do users show improvement through tasks with time, navigation, and error rates? Do users show more or less frustration while completing tasks?
- Satisfaction: Verbal, observation, questionnaire, or interview data. Do users indicate satisfaction with the product? Do users seem frustrated, interested, content, etc., when using the product?

It may be that you are only interested in efficiency and learnability for one iteration of your test. In this case, you can combine methods for MEELS and *see-say-do* to decide which methods would best suit your needs. Essentially, you should adjust your data collection methods based on your specific research questions (Box 8.8).

It is important that you plan in advance how you will record your data, and this should be determined based on your environment. What technologies and tools will you have access to? What technologies and tools are appropriate (e.g., nondistracting) for the testing situation? If you are testing a website or software, you may want to use usability testing and analysis software such as TechSmith's Morae (for Windows) or Silverback (Mac). You can also create your own usability testing equipment by using a screen capture and audio capture tool (see Tomlin 2014 for other options). If you are testing in an environment with no computers, you may consider using audio and visual recording devices to capture data. You still need to take observation notes during these sessions to ensure that data is recorded in case of technology mishaps, but it is useful to have visual and audio captures so you can refer back to the testing session. When it is not possible to record sessions at all, you can still record data using observation and recording sheets (see "sample data collection sheet" at the end of this chapter).

BOX 8.8   DATA COLLECTION FOR EYEGUIDE FOCUS PROTOTYPE TEST (COURTESY OF EYEGUIDE)

- User interaction with the prototype will be collected in three primary categories: observational, verbal, and performative. All of this data will be recorded by both the moderator and computer.
- Observational data will be noted by experimenters throughout the course of the test. These notes will be collected on the data sheet and observations may include notes about lapses in the system's functionality, user's mental models expressed through action or verbalization, or anything else deemed interesting by the researchers.
- Verbal data will be offered by the participants throughout the session and recorded by the experimenters. The protocol describes the think-aloud method we'll ask participants to use as well as the opportunities for active intervention the experimenters may take advantage of. If the participant falls silent for three seconds or more, the moderator will prompt the participant to continue by asking, "What are you thinking?" By collecting the participant's thoughts as they work with the prototype and pursuing clarification or explanation when appropriate, we hope to come to a better understanding of how the user intends to use the system as well as the problems they encountered while doing so.
- Additionally, verbal data will be collected once all tasks are completed; the experimenters will ask some prepared questions about the design and some impromptu questions about the task session. The participant will be encouraged to continue talking until they have no additional comments about the prototype.
- Observational and verbal data will both be recorded on data sheets. The experimenters will make an on-site categorization judgment of negative, neutral, or positive based on whether the item hinders, doesn't affect, or helps the user toward his or her goal.
- Severity rankings will be added post hoc, performed by each experimenter independently on his or her own observational sheet. Severity rankings will be based on a modified form of the Dumas–Redish scale adjusted to comment on both the positive and negative aspects of the system:
- Something occurs that prevents/makes impossible the completion of the task
- Something happens that creates/negates delay and frustration
- Something has a minor positive/negative effect on usability
- There is just a subtle issue
- The table below lists the performative data to be collected during each task. Both experiments will record performance on their individual data sheets

8.3.7 DATA ANALYSIS

When planning your test, you also want to decide how you will analyze your data. This not only helps the analysis go faster, but also informs how you choose to collect data and create recording sheets. You should not analyze data while testing is in progress. Analyzing while testing can lead to biased assumptions about how testing is going, which can then influence your analyses of later test sessions. You should, however, mark observation events during the test, but wait to analyze or code these observations until testing is complete. Further, for qualitative data that requires additional coding (TAP, active intervention, interviews, errors, etc.), you need to have more than one person on your team analyze the data to ensure accuracy.

Qualitative analysis (first mentioned in Chapter 3) is appropriate for data that cannot be counted (i.e., quantitative data). Qualitative data must be interpreted for meaning. You will most likely encounter two types of qualitative data: (1) transcripts from TAP, active intervention, or interviews and (2) error coding. For each of these analyses, you need to code data using either a top-down or bottom-up coding method. Top-down coding means that you already have a coding criteria or heuristic that you will use to interpret data. Bottom-up coding, sometimes called emergent coding, means you look for patterns that emerge from the data. Using these patterns, you categorize data into descriptive groups that help you understand how users interacted with the product. If you already know what type of feedback you are looking for, top-down coding may be the most effective analysis tool. However, if your test is more general and discovery in nature, then bottom-up coding would reveal the most new data from that test. Thus, choose coding processes that best fit the purposes of the test.

Errors are often coded using "severity" scales. Several scales can be used or adapted, such as the Dumas–Redish scale (see Box 8.8 for an example). Dumas and Redish (1999) break down error severity into four levels:

1. Level 1: Problems prevent completion of a task.
2. Level 2: Problems create significant delay and frustration.
3. Level 3: Problems have a minor effect on usability.
4. Level 4: Problems are more subtle and often point to an enhancement that can be added in the future (324).

Several other scales can be used or adjusted based on your testing needs. Sauro (2013) lists these scales from Jakob Nielsen, Jeff Rubin, Chauncey Wilson, and Molich and Jeffires as well as his own approach. All are similar scales, but their terminology and degree of severity may differ.

Quantitative analysis should be used on data that can be counted, such as time on task, mouse clicks, eye tracking, and multiple choice or Likert surveys. Quantitative data (also first mentioned in Chapter 3) are usually expressed numerically and include statistical analysis when running A/B or summative tests (see Sauro and Lewis 2012 for more information on statistical analysis). Eye tracking is the exception to numerical reporting. Eye tracking is a quantitative method, but you may use more descriptive analyses that illustrate where users are looking (like cluster areas), where the most activity is seen on a website (heat map analysis), or the visual navigation of the users (gaze plots). In each of these cases, numerical coordinates are available for reporting and cross-checking, but you will most likely look for "where" users are looking in relation to other areas of the product.

8.4 CONDUCTING YOUR TEST

When conducting your test, you'll need to consider and prepare for several points, including interacting with users, writing a test script, and making error and observation notes while testing. After you have a script and plan, you should always perform pilot testing to be sure participants understand the scenario and tasks, that test procedure goes smoothly within the usability team, and that the test environment is suitable for your test.

8.4.1 User Interaction

Testing seems easy enough, but there are considerations for preparing for and interacting with your users. One of the first things you need to consider is what kind of legal or organizational procedures you need to follow. Your organization should have research resources you can consult. If working in or for a public institution or university, you will probably need approval from the human research board before you begin testing; this can be a lengthy process, so apply for approval when you first begin your research study. At minimum, you should have a participant consent form for testing. A larger corporation probably has a boilerplate consent form you can use.

In addition to getting permission to research, you need to prepare your participants for testing. When participants hear "test," they assume there is a way to fail said test. From the beginning of your interaction with users (even when you send emails for recruitment), you need to assure users that they are not being tested—the *design* is being tested. This means explaining clearly to users what is being tested, why the product is being tested, what is being asked of them, and how the "testing" process will proceed. Although facilitators are told to be neutral, Kate often tells her participants, "There is no way you can fail this test. It is the best test you'll ever take." Putting users at ease will help them focus on the tasks at hand.

8.4.2 Test Script

Once you have created a test plan, you need to create a testing procedure script that your team will use to conduct every test session. The script is a detailed map of everything that should happen from setting up the room to engaging with the user and even saving data. When writing a script, imagine what a movie script looks like. These scripts describe settings, actor lines and actions, lighting, music, sounds, etc. You don't have to worry about mood music, obviously, but you want to script out the interaction in as much detail as possible (see Box 8.9 for a checklist and the "sample paper prototype script" at the end of the chapter).

BOX 8.9   TEST SCRIPT CHECKLIST

TEST SCRIPT CHECKLIST

- ❑ M.aterials for testing
- ❑ Test environment setup
- ❑ Team roles
- ❑ Script
  - Test background
  - Consent form
  - Facilitator lines during test
  - Reminders for TAP, active intervention questions
  - Posttest interview or questionnaire, etc.
  - End of test indication
  - Thank user for participating
- ❑ Protocol for saving data

When writing a script, you should keep a few key points in mind. First, the script should be written in enough detail that if, for some reason, one or more people is missing from the team, the test can still proceed. This means the test environment is clearly described. Box 8.10, for instance, clearly outlines the materials needed for the test, the furniture's position, and how the materials will be used.

The facilitator's speech should be written word for word as it will be spoken during the tests. It is important, especially when testing multiple users, to be consistent with what you say so that one user is not given more or less information to complete tasks than others. When writing the script, you should first provide background information about the test (Box 8.11). Explain what the product is, how the product is used, and what scenario the user will test under. At this point, make sure to have the user sign a consent form for testing (if applicable).

When explaining the test, make sure users understand that the design is being tested, not their personal performance. The goal of testing is to see how well the product holds up to use, not how well users perform to designers' expectations.

The script should then break down what the facilitator should say and what actions participants will be asked to take. Often, the test begins by providing users with a test scenario, which is a fictional yet representative context under which users will test (Box 8.12). From there, each task is explained as well as the actions users should take to indicate when tasks are finished.

Finally, you need to include the posttest activity in the script; this includes interview questions or questionnaires (Box 8.13).

This may seem like a lot of information to document. However, the more specific you are, the more prepared you are, especially if, again, one of your team members is absent. Because recruiting participants is hard enough, you want to make sure your

**BOX 8.10   TEST ENVIRONMENT DESCRIPTION
IN TEST SCRIPT (COURTESY OF EYEGUIDE)**

Materials brought to the test:

- Three black or blue pens
- Script (pages 2–7 of this appendix)
- Data sheets (Appendix I)
- Task and scenario papers (Appendix H)
- Prototype
- iPad (with Focus app and "Demo" profile created)
- Eye-tracking equipment
- Something to prop the iPad up with
- Postexperimental questionnaire

When we initially arrive to the testing area, we will exchange pleasantries with the participant, then ask that the participant step out while we set up the testing. We'll also give the participant the consent form to look over and sign while we set up. We'll lay out the furniture so that it approximates the following picture:

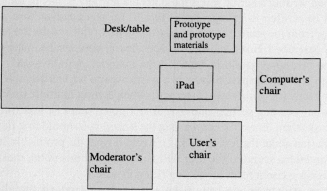

The prototype materials are the pieces of the prototype that the computer manipulates when the user interacts with the prototype. Computer will place prototype and prototype materials the way he or she wants them. Moderator will turn on iPad, pull up Focus, then ensure all paper testing materials are in order.

**BOX 8.11   TEST INFORMATION IN TEST
SCRIPT (COURTESY OF EYEGUIDE)**

As mentioned in the consent form, the goal of this study is to test out a new concussion detection product made by a company called EyeGuide. It's important to remember that we're testing the product, but we are not testing you. You are here to help us understand how trainers may use a product like this. We are interested in what is working well with the current design and what should be improved. Accordingly, we need and appreciate your honest feedback.

The product is an iPad app called EyeGuide Focus. It uses eye tracking to detect and monitor concussions. The basic idea is that when someone gets a concussion, his or her visual attention suffers. As a result, if he or she tries to follow a moving dot on a screen, the eye movements are less smooth than those of someone without a concussion. We ask student athletes to look at a dot on the iPad screen and follow it using just their eyes. EyeGuide Focus will track the smoothness of the athlete's eye movements.

BOX 8.12   TEST SCRIPT SCENARIO (COURTESY OF EYEGUIDE)

The iPad program in front of you has the information of athlete Taylor Hutchens pulled up.

Taylor was diagnosed with a concussion last week. Since then, you have given Taylor the eye-tracking test several times. The EyeGuide Focus app has all of the eye-tracking tests stored, including a preinjury baseline test the athlete took at the beginning of the school year. Additionally, the global baseline data, the averages for all athletes, is stored in the system.

[Set down Task 1 paper.]

Task 1: You remember that Test 1 was the baseline test the athlete took at the beginning of the school year. Indicate to the EyeGuide Focus program that Test 1 is the baseline. When you have finished, please say to me, "I've finished."

test is prepared when they come; there is no guarantee they would come back to test again if you are unable to test during their initial appointment.

8.4.3   ERROR/OBSERVATION NOTES

You should have a clear plan about how your observers will take notes during the test. You should have a system for documenting this information in your test plan and follow it during testing. However, you should not interfere with users as they work. In fact, you want the fewest number of people in the room to eliminate the Hawthorne effect or the chances that your users will overperform because they are being watched (Macefield 2007). Therefore, decide who needs to be in the room and what information they should be collecting.

**BOX 8.13   POSTTEST ACTIVITY IN TEST
SCRIPT (COURTESY OF EYEGUIDE)**

*[Computer cleans up prototypes, puts them in their folder. Moderator pulls
out posttest questionnaire.]*
   This questionnaire will ask you about your experience with this design of
EyeGuide Focus and about yourself. Your answers to these questions will be
confidential. [Computer] and I will be working on something else while you
complete this questionnaire, so take as much time as you need. Let me know
when you've finished.
   *[While participant starts the questionnaire, moderator and computer work
on filling in empty severity ratings in the observation sheet. They will do this
on their individual clipboards and slightly removed from the table to give the
participant privacy and to prevent participant from viewing the observation
sheets. When participant has finished questionnaire, Moderator puts it in a
manila folder without looking at it.]*
   Thank you. Do you have any other comments about this design of the
EyeGuide product?
   [If participant gives an answer:]
   Anything else?
   [Continue asking "Anything else" until participant says no.]

8.5   ANALYZING YOUR DATA

In the test plan section, we discussed the importance of understanding how you
would analyze data. Here, after the test, is where you actually analyze data. Do not
analyze data before the test is over. Coding of and compiling data should be saved
for after the test to avoid potential bias.
   Much of your quantitative data will be averaged by task and provide, at least, the
standard deviation. If testing only a small number of participants (five to 10) and
measuring only time on task, dwell time, and mouse clicks, you may not need much
else. However, if conducting a summative or A/B test, further analysis is needed (see
Sauro and Lewis 2012). To code errors, you can find the average error rate based on
severity and provide the frequency of each error type by task. Errors should be coded
by at least two people to ensure accuracy.
   To analyze your qualitative analysis, such as think-aloud protocol or posttest
interviews, you should look for patterns in feedback. The "sample data collection
sheet" at the end of this chapter breaks down observation types by positive, neutral,
or negative. This is one way to look for patterns. Another way is to use emergent
coding with which you look for similar statements that discuss problems, confusion,
error, etc., and compile those comments together. This, then, gives more specific
information about usability issues that users observe.
   If testing a small group, you may have an outlier; this is a test participant's results
that are extremely different from other users. You need to decide if this outlier truly
represents a usability problem or whether the outlier should be thrown out. To make

this decision, Barnum (2011) recommends reexamining the user in relation to the representative population, determining if others would have the same problem, and deciding if the problem needs to be investigated further. If the user was not part of the representative population, then the outlier could be thrown out. However, if the outlier's issue is a problem from a representative user that could be fixed, then it should be reported to the designers and fixed.

When you are ready to synthesize the data for your findings, you want to bring the *see-say-do triangle* together. How do your observation, feedback, and performance data come together? Are there patterns emerging from this data? If so, what do the patterns say about the design at this time?

Let's look at an abbreviated example of how all this works. When a hi-fi prototype of EyeGuide Focus was tested with real users, developers had a number of testing goals. One of these centered on concerns, raised in previous development and testing, about frustration users experienced when adding new players to the system. Developers knew that a lot of demographic information (e.g., athlete's name, age, height, weight, sports played) needed to be added to complete a player's profile. But how much was too much? How long did it take to do this, and could this time be shortened? Did all the steps for creating a profile introduce error?

To determine the MEELS of adding a new player to Focus, developers asked representative users first to add a complete player before selecting that player for testing. In all, five users participated in testing, and the results were enlightening. For one, the average user took more than three minutes to add a new player. If an athletic trainer had to add 100 users every year, all representing new athletes in an average-sized high school sports program, this would represent roughly five hours of data entry, on an iPad, every year. Beyond the time it took to do this, developers also learned that athletic trainers made errors. Four out of the five athletic trainers made mistakes entering birthdates. Plus, many didn't know all of the players' information, such as student IDs or all sports played. Because the process was long and the scenario required users to add players before advancing, developers observed (See) that three of the five users elected to rush through the profile, saving most of the data entry for later so they could move on to what they regarded as most important, which was baseline testing the athletes.

Users complained (Say), at the conclusion of testing, about the length and difficulty of creating player profiles using an iPad, and this information, when triangulated against poor performance data (Do) and what the developers observed, led the EyeGuide team to conclude that the process for adding players' demographic data needed improvement. As a result, a future design of the Focus prototype allowed athletic trainers to skip adding the detailed profile when creating a baseline, opting instead for a shortcut that required just the athlete's name. Athletes were marked in the searchable database as needing a completed profile, and this worked well because a system was already in place for marking athletes who had not yet been baselined or who were concussed and in recovery and monitoring. Further—and this speaks very much to the value of usability testing as a return on investment (ROI) not just for the existing prototype as it is but also for what it can be—developers came up with a way to migrate a spreadsheet of athletes with all of their information directly into Focus so that most of the athletes' profiles were created en masse, quickly, without requiring manual input from the athletic trainers. Because such a feature was a value-add, it was turned into an add-on that schools could pay extra for to use, making profile creation easier and more usable for them and generating extra revenue for EyeGuide.

Voluminous data results, from hundreds of users, aren't necessary to make a product better for its intended users. One simple task with five users in a realistic scenario, as demonstrated above, aided EyeGuide in understanding what was wrong and how it could be solved. Users performed, gave feedback, and were observed (see-say-do), and the results helped to create a better, more user-friendly, and more revenue-friendly, product.

8.6 WHY YOU NEED USABILITY TESTING

As discussed in Chapter 7, the Lean UX approach advocates building and testing a prototype quickly so that users can be incorporated constructively into a typically agile or fast moving (sprint) design process. In this chapter, we've provided you with a template, along with examples and explanations, so that you know the right way to construct a formal, valid, and rigorous usability test as part of your UCD efforts.

But if you don't have time, if you are limited in access to users, or if you are short on money or help from others, you can take what we've described here and reduce its scope and formality. The most important thing is that you have users use your product in realistic situations and that you gather feedback from them, adhering to the *see-say-do triangle*, to include in future iterations of the product.

At EyeGuide, the tendency is to develop a more elaborate testing plan at the beginning to ensure that everything is covered for a sound testing process. From there, however, to save on time so that EyeGuide developers can keep the design process moving forward as quickly and effectively as possible, test planning and implementation are often shortened in time and narrowed in scope. Developers still have test goals, make clear what they will test and what data they intend to collect, and have a script, but because they already know the overall strategy for testing because they've previously built that out, they implement more focused, faster testing so that they can get feedback and integrate it as soon as possible.

For example, when EyeGuide deployed an MVP of its concussion results display screen (see Chapter 7), representative users were already in the audience. Rather than adhere to a formal testing process that might take hours to set up and even more to implement, developers took a more lean approach. They simplified what they wanted from the users. How did they react to the concussion results display? Was there any confusion, concern, did they appear to like or dislike it (See)? What did they say (Say)? What did they do or want to do with it (Do)? Developers already had the right population, and the scenario was obvious (athletic trainers testing athletes for a concussion), so they pivoted quickly from presenting an MVP of the results screen to gathering feedback from real users about that screen. Formal test planning, as described in this chapter, allowed them to have fundamentals in place to apply them in a more focused way.

Don't ever argue that usability testing is too complicated or too long of a process to implement as an excuse for not involving users. You can reduce how long it takes to carry out. As long as you test real users doing real things with your product in real situations, you will conduct worthwhile usability that aids your design. If you don't do it, even in a limited way, how do you know what your users really think, say, and do with your product?

TAKEAWAYS

1. You should test your product (from prototype to completed versions) frequently with small populations of representative users each time. This is known as iterative testing, meaning that you test the design *throughout* its development.
2. Design your test around your test and design goals. Choose testing methods that will best inform that stage of design.
3. The mantra of usability testing is that you test representative users performing representative tasks.
4. A/B testing and formative testing are the two most useful tests for comparing and testing products during your design process.
5. Participants need to be representative users, and the sample size needs to be appropriate for the type of test you run.
6. Write a detailed test plan and test script to ensure you and your team are prepared to test.
7. Using real user data in usability testing rather than data given to users to use during the test can increase user investment, reduce cognitive load, and ensure realistic data is used.
8. Cognitive overload can occur when users are asked to do too much, and it may affect their ability to perform tasks due to multiple outside influences tapping their cognitive resources.
9. Use MEELS to determine the usability of your product.
10. Don't analyze data while testing is in progress; this can lead to biased assumptions about how testing is going and influence your analysis for later test sessions.
11. Putting users at ease will help them focus on the tasks at hand—let them know that the *product* is being tested, not them.
12. The facilitator's speech should be written—word for word—as it will be spoken during the tests. Be consistent with what you will say so that one user is not given more or less information to complete tasks than others.
13. Voluminous data results from hundreds of users aren't necessary to make a product better for its intended users. One simple task with five users in a realistic scenario can aid in understanding what is wrong and what can be done as a solution.

DISCUSSION QUESTIONS

1. Think about the efficiency component of the MEELS process. What information might you need to determine what a "reasonable" amount of time is for users to complete tasks? What if some users complete tasks in a very short or very long amount of time, but they are ultimately both satisfied? Do you have a successful product?
2. When might you decide to use summative or formative testing? What are the benefits and drawbacks of each approach?
3. What types of product components would be useful to test in an A/B test? Can you come up with A/B test components for a clothing website, cell phone company website, and mobile music app?

4. Can paying money to test participants sway test results? Does it matter as long as they are representative users? How else might you incentivize users to participate?
5. What can you do to make the test environment as representative as possible?

REFERENCES

Barnum, C. *Usability Testing Essentials: Ready, Set...Test!* Boston: Morgan Kaufmann, 2011.

Bergstrom, J., E. Olmstead-Hawala, J. Chen, and E. Murphy. "Conducting Iterative Usability Testing on a Web Site: Challenges and Benefits." *Journal of Usability Studies* 7, no. 1 (2011): 9–30.

Caulton, D. "Relaxing the Homogeneity Assumption in Usability Testing." *Behaviour & Information Technology* 20, no. 1 (2001): 1–7.

Cooke, L. "Eye Tracking: How It Works and How It Relates to Usability." *Technical Communication* 52, no. 4 (2005): 456–463.

Duchowski, A. *Eye Tracking Methodology: Theory and Practice.* London: Springer, 2007.

Dumas, J., and J. Redish. *A Practical Guide to Usability Testing.* Portland: Intellect, 1999.

Faulkner, L. "Beyond the Five-User Assumption: Benefits of Increased Sample Sizes in Usability Testing." *Behavior Research Methods, Instruments, & Computers* 35, no. 5 (2003): 379–383.

Genov, A., M. Keavney, and T. Zazelenchuk. "Usability Testing with Real Data." *Journal of Usability Studies* 42, no. 2 (2009): 85–92.

Gothelf, J. *Lean UX: Applying Lean Principles to Improve User Experience.* Sebastopol: O'Reilly Media, 2013. Kindle Edition.

Hawley, M. "Modifying Your Usability Testing Methods to Get Early-Stage Design Feedback." *UXmatters.* July 24, 2012, accessed February 22, 2016. http://www.uxmatters.com/mt/archives/2012/07/modifying-your-usability-testing-methods-to-get-early-stage-design-feedback.php.

Macefield, R. "How to Specify the Participant Group Size for Usability Studies: A Practitioner's Guide." *Journal of Usability Studies* 5, no. 1 (2009): 34–45.

———"Usability Studies and The Hawthorne Effect." *Journal of Usability Studies* 2, no. 3 (2007): 145–154.

Morgan, M., and L. Borns. "360 Degrees of Usability." Proceedings of the Conference on Human Factors in Computing Systems, Vienna, Austria, April 24–29, 2004: 795–809.

Nielsen, J. 2012a. "Usability 101: Introduction to Usability." *Nielsen Norman Group.* January 4, 2012, accessed February 22, 2016. https://www.nngroup.com/articles/usability-101-introduction-to-usability/.

———. 2012b. "A/B Testing, Usability Engineering, and Radical Innovation: What Pays Best?" *Nielsen Norman Group.* March 26, 2012, accessed February 22, 2016. https://www.nngroup.com/articles/ab-testing-usability engineering/.

———. 2012c. "How Many Test Users in a Usability Test?" *Nielsen Norman Group.* June 4, 2012, accessed February 22, 2016. https://www.nngroup.com/articles/how-many-test-users/.

———. 1993. *Usability Engineering.* San Francisco: Morgan Kaufmann.

Ramey, J., T. Boren, E. Cuddihy, J. Dumas, Z. Guan, M. Van den Haak, and M. De Jong. "Does Think Aloud Work?: How Do We Know?" Proceedings of the Association for Computing Machinery Conference on Human Factors in Computing Systems, Extended Abstracts on Human Factors in Computing, Montreal, Canada, April 22–27, 2006: 45–48.

Russell, D., and M. Oren. "Retrospective Cued Recall: A Method for Accurately Recalling Previous User Behaviors." Proceedings of the 42nd Hawaii International Conference on System Sciences, Hawai'i, Hawai'i, January 5–8, 2009: 1–9.

Sauro, J., and J. Lewis. *Quantifying the User Experience: Practical Statistics for User Research*. Boston: Morgan Kaufmann, 2012.

Sauro, J. "Rating the Severity of Usability Problems." *MeasuringU*. July 30, 2013, accessed February 22, 2016. http://www.measuringu.com/blog/rating-severity.php.

———. "7 S's of User Research Sampling." *MeasuringU*. March 13, 2012, accessed February 22, 2016. https://www.measuringu.com/blog/sampling-s.php.

Spool, J., and W. Schroeder. "Testing Web Sites: Five Users Is Nowhere Near Enough." Proceedings of the Association for Computing Machinery Conference on Human Factors in Computing Systems, Extended Abstracts on Human Factors in Computing, Seattle, Washington, March 31–April 5, 2001: 285–286.

Tomlin, W. "14 Usability Testing Tools Matrix and Comprehensive Reviews." Useful Usability. February 10, 2014, accessed February 22, 2016. http://www.usefulusability .com/14-usability-testing-tools-matrix-and-comprehensive-reviews/.

SAMPLE DATA COLLECTION SHEET
(COURTESY OF EYEGUIDE)

Task 1: You remember that Test 1 was the baseline test the athlete took at the beginning of the school year. Indicate to the EyeGuide Focus program that Test 1 is the baseline. When you have finished, please say to me, "I've finished."

Correct click path: Completed task?   yes   no
❑   Tests
❑   Edit
❑   Mark as baseline
❑   Save

Observation/Verbalization	Type			Severity (Dumas-Redish Scale)			
	Positive	Neutral	Negative	4	3	2	1

SAMPLE PAPER PROTOTYPE SCRIPT
(COURTESY OF EYEGUIDE)

Note: We are visiting users at their place of work. We expect, based on what was agreed upon in the phone conversation, that the user will provide a secluded room or work area with a table or desk and three chairs.

Materials brought to the test:

- Three black or blue pens
- Script (pages 2–7 of this appendix)
- Data sheets
- Task and scenario papers
- Prototype
- iPad (with Focus app and "Demo" profile created)
- Eye-tracking equipment
- Something to prop the iPad up with
- Postexperimental questionnaire

When we initially arrive to the testing area, we will exchange pleasantries with the participant, then ask that the participant step out while we set up the testing. We'll also give the participant the consent form to look over and sign while we set up. We'll lay out the furniture so that it approximates the following picture:

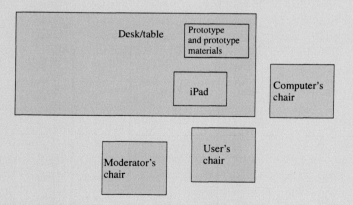

The prototype materials are the pieces of the prototype that the computer manipulates when the user interacts with the prototype. Computer will place prototype and prototype materials the way he or she want them. Moderator will turn on iPad, pull up Focus, then ensure all paper testing materials are in order.

Once we have the test set up, we will invite the user back in and initiate the following script.

Script

[Computer is sitting. Moderator stands in doorway to invite in and greet the user.]

Hi! Thanks for having us and for participating in our study. My name's [moderator's name], and I'll be walking you through what we're doing today and answering any questions you have. This is [computer's name], who will be here to assist with the study. I'd like to tell you more about what we'll be doing today, but first will you please sit down in this middle chair?

[Moderator and user sit down.]

First, I want to apologize for reading off of this paper. I'm doing this to ensure that I don't forget to tell you anything and so that I can be sure I say more or less the same thing to all participants in this study.

Do you have any questions about the consent form we gave you?

[Answer questions. If the user has not yet signed the consent form, ask: "Are you ready to sign the consent form and begin our study?" *After the user has signed the consent form and is finished asking questions, resume the script.]*

As mentioned in the consent form, the goal of this study is to test out a new concussion detection product made by a company called EyeGuide. It's important to remember that we're testing the product, but we are not testing you. You are here to help us understand how trainers may use a product like this. We are interested in what is working well with the current design and what should be improved. Accordingly, we need and appreciate your honest feedback.

The product is an iPad app called EyeGuide Focus. It uses eye tracking to detect and monitor concussions. The basic idea is that when someone gets a concussion, his or her visual attention suffers. As a result, if he or she tries to follow a moving dot on a screen, the eye movements are less smooth than those of someone without a concussion. We ask student athletes to look at a dot on the iPad screen and follow it using just their eyes. EyeGuide Focus will track the smoothness of the athlete's eye movements.

To give you a better idea of what I'm talking about, I'm going to demonstrate how you and other athletic trainers might give a student athlete the eye-tracking test. I'll pretend to be a trainer, and [computer] will be the student athlete. First, I'll set up the test on the iPad.

[Moderator turns on iPad, brings it to list of athletes.]

I select the athlete's profile...

[Moderator pulls up "Demo" profile]

...and now I should get the equipment ready for the test.

[props iPad up. Moderator stands.]

[to computer] [Computer], will you please come sit here and put on the eye-tracking equipment.

[Computer sits in Moderator's chair, puts on the headphones.]

[to user] The eye-tracking equipment that will actually be used with EyeGuide Focus hasn't been built yet. This is the best we have for the time being.

Now let's calibrate the camera.

[presses "Run Test" button]

This step is to make sure that the camera can actually see the athlete's eye.

[Moderator adjusts camera, computer continually looks at four corners of the iPad.]

That looks pretty good. Now let's begin the test. [Computer], remember to keep your head still.

[Press "Run Test." When over, Moderator continues.]

Okay. Let's see the replay. *[to participant]* This replay will show a red dot in addition to the white dot we just saw. The red dot indicates where [Computer] was looking during the test.

[Press "View Replay." When over, Moderator continues.]

[to participant] You can see how the program has recorded the information about where the athlete was looking in relation to where [he or she] was supposed to look. Athletes without concussions will generally be better at following the white dot than those without concussions. Their eye movements will have less variability than those with concussions.

[While Moderator says the following, Computer removes headphones, puts away iPad and eye tracker, returns to original seat and sets up paper prototype. Moderator sits back down in original seat.]

So far, there is no way to look at how well the athlete did on the eye-tracker test other than watching the replay video. EyeGuide has started designing other ways to display the eye-tracker information. You'll help us get an idea of what is good about the design and whether it has any problems.

As I've mentioned, the design is in the very early stages. Right now, we just have a version that is made out of paper. You can still use it like you would an iPad app; [Computer] will be playing computer. When you press a button, she will update the screen.

We're going to give you a few tasks to perform on this application. Again, we are not testing you. The tasks are designed to test the current design.

I will read each new task aloud and set a piece of paper down here on the table for you to refer to. If you have any questions about a task you may ask me. However, I will not be able to tell you how to complete the task. Anyway, sometimes there is no single right answer or way to complete the task.

While you're doing the tasks, please talk about what you are doing and why. [Computer] and I will take notes on what you are saying. If you fall silent I will prompt you to keep talking.

Before we begin, do you have any questions?

[Answers questions, when applicable. When ready to begin, set down scenario paper.]

The iPad program in front of you has the information of athlete Taylor Hutchens pulled up.

Taylor was diagnosed with a concussion last week. Since then, you have given Taylor the eye-tracking test several times. The EyeGuide Focus app has all of the eye-tracking tests stored, including a preinjury baseline test the athlete took at the beginning of the school year. Additionally, the global baseline data, the averages for all athletes, are stored in the system.

[Set down Task 1 paper.]

Task 1: You remember that Test 1 was the baseline test the athlete took at the beginning of the school year. Indicate to the EyeGuide Focus program that Test 1 is the baseline. When you have finished, please say to me, "I've finished."

[When finished, pick up Task 1 paper and set down Task 2 paper.]

Task 2: Ultimately, you would like to decide today whether Taylor should continue rehabilitating or whether you can recommend that he see a physician and resume play. Select the tests you would like to compare in order to make that decision. When you have finished, please say to me, "I've finished."

[When finished, pick up Task 2 paper and set down Task 3 paper.]

Task 3: With the information stored in the system, decide whether Taylor should continue rehabilitating or whether you can recommend that he see a physician and resume play. When you have made your decision, please tell me either that he should continue rehab or see a physician.

[When decision is made, pick up Task 3 paper and set down Task 4 paper.]

Task 4: Rename one of the eye-tracking tests. When you have finished, please say to me, "I've finished."

[When finished, pick up Task 4 paper and scenario 1 paper. Set down Scenario 2 paper (which also has Task 5 on it). Computer picks up prototype, replaces the relevant parts with Athlete 2's information, and sets the prototype back in front of the participant.]

Now you have the information of athlete Steven Mendoza up on the iPad. Steven was also diagnosed with a concussion last week. Since then, you have given Steven the eye-tracking test several times. The EyeGuide Focus app has all of the eye-tracking tests stored. You'll notice that this time, the first test wasn't taken until after the injury—Steven has no preinjury baseline. As before, the global baseline data, the averages for all athletes, are stored in the system.

Task 5: With the information stored in the system, decide whether Steven should continue rehabilitating or whether you can recommend that he see a physician and resume play. When you have made your decision, please tell me either that he should continue rehab or see a physician.

[When decision is made, pick up Scenario 2 paper. Computer arranges the results pages of both athletes so that they are next to each other in front of the participant.]

Thank you for your participation so far. We have concluded the part of the study where you complete tasks. Now I'm going to ask you a few questions about the displays you see in front of you. Again, we are not testing you; we are testing the displays. We need to know which parts are helpful and easy to understand and which parts aren't, so please be honest and thorough when you answer the questions.

[The following are three prepared questions, the third of which also has two follow-up questions. It is appropriate for the moderator to ad lib questions at this point.]

What composite score would indicate to you that an athlete is not suffering from a concussion?

Why do higher scores indicate the presence of a concussion and lower scores indicate the lack of a concussion?

If I was a student athlete and asked you what the variability of eye-movement graphics meant, what would you tell me?

- What do the colors [in the variability of eye-movement graphic] mean?
- What do you learn by comparing a baseline variability graphic to a postinjury test graphic?

[Computer], do you have any additional questions?

[Once Moderator and Computer have exhausted their questions:]

Ok, let's move on.

[Computer cleans up prototypes, puts them in their folder. Moderator pulls out posttest questionnaire.]

This questionnaire will ask you about your experience with this design of EyeGuide Focus and about yourself. Your answers to these questions will be confidential. [Computer] and I will be working on something else while you complete this questionnaire, so take as much time as you need. Let me know when you've finished.

[While participant starts the questionnaire, moderator and computer work on filling in empty severity ratings in the observation sheet. They will do this on their individual clipboards and slightly removed from the table to give the participant privacy and to prevent participant from viewing the observation sheets. When participant has finished questionnaire, Moderator puts it in a manila folder without looking at it.]

Thank you. Do you have any other comments about this design of the EyeGuide product?

[If participant gives an answer:]

Anything else?

[Continue asking "Anything else" until participant says no.]

Again, thank you for taking the time to be in our study. Would you like any help putting the furniture back to the way it was?

Scenarios and Tasks to Hand to Participant

The iPad program in front of you has the information of athlete Taylor Hutchens pulled up.

Taylor was diagnosed with a concussion last week. Since then, you have given Taylor the eye-tracking test several times. The EyeGuide Focus app has all of the eye-tracking tests stored, including a preinjury baseline test the athlete took at the beginning of the school year. Additionally, the global baseline data, the averages for all athletes, are stored in the system.

> Task 1: You remember that Test 1 was the baseline test the athlete took at the beginning of the school year. Indicate to the EyeGuide Focus program that Test 1 is the baseline. When you have finished, please say to me, "I've finished."
> Task 2: Ultimately, you would like to decide today whether Taylor should continue rehabilitating or whether you can recommend that he see a physician and resume play. Select the tests you would like to compare in order to make that decision. When you have finished, please say to me, "I've finished."
> Task 3: With the information stored in the system, decide whether Taylor should continue rehabilitating or whether you can recommend that he see a physician and resume play. When you have made your decision, please tell me either that he should continue rehab or see a physician.
> Task 4: Rename one of the eye-tracking tests. When you have finished, please say to me, "I've finished."

Now you have the information of athlete Steven Mendoza up on the iPad. Steven was also diagnosed with a concussion last week. Since then, you have given Steven the eye-tracking test several times. The EyeGuide Focus app has all of the eye-tracking tests stored. You'll notice that this time, the first test wasn't taken until after the injury—Steven has no preinjury baseline. As before, the global baseline data, the averages for all athletes, are stored in the system.

> Task 5: With the information stored in the system, decide whether Steven should continue rehabilitating or whether you can recommend that he see a physician and resume play. When you have made your decision, please tell me either that he should continue rehab or see a physician.

9 RIDE (Report, Iterate, Deploy, Evaluate)

You've done a lot of work to this point. You've made it all the way through the RABBIT process for effective UCD. You've researched the users of your product, assessed the situation(s) in which they'll use that product, and even created an operative image and tested it. Given how much you've done and how careful you've been to include and design for users throughout the process, you should feel pretty good.

So take a moment to congratulate yourself if you're doing this on your own, or thank and applaud members of your team who have helped you throughout to build from the ground up a truly UCD product. Just remember, however, one important thing in the midst of your celebration: You're not done yet.

In fact, you're not really ever done. This can be a challenging and frustrating concept for many of you who are goal-oriented and like to feel as if once you've completed a project you can move on to something else. But UCD doesn't work like that. Users change, the situations in which they use the product change, the materials of the product change (the smartphone screen gets larger, the Internet gets faster, etc.), and so on. Fundamentally, UCD is a recursive process, one that lends itself to proactive monitoring and ongoing improvement of a product to meet user needs and wants. Should the product be less dynamic with a life cycle that doesn't require it to be updated because of other limitations (such as a microwave oven that users may not replace for years), you may not constantly work to update its control interface to make it more design-friendly for its users. However, you will still gather feedback from users, and you will use that knowledge to design other products they use.

To aid you with what happens after you have made and tested your UCD product, we introduce another acronym: RIDE. It isn't enough to make a product for users and test it. You must do the following:

- *Report* what you've learned from testing to improve that product or to create a repository of information that can be used for other products.
- *Iterate* the RABBIT process to collect new user and situation of use information and modify products (through prototyping and testing) by incorporating what you've learned.
- *Deploy* the newly revised product.
- *Evaluate* that product again with user feedback to improve it, just as you evaluate the overall RABBIT process to improve user and situational knowledge.

Think of RABBIT as what you do within the parameters of the product design project to create a UCD product. Once a product has been created using RABBIT, think of RIDE as what you do to ensure you are (1) doing what is necessary to inform you, your team, or others about your UCD projects and also (2) implementing an approach for design that enables constructive, iterative modification and evaluation of your projects' products.

In this chapter, we will cover RIDE's key parts in detail, offering best practices and examples for how to carry it out.

9.1   REPORT

Just as a rough paper prototype of the eventual final product serves as an operative image that users can interact with to give you feedback, a report, even short and informal, serves in much the same way. A report presents results of user interactions with your product. You need an operative image because it transfers designer vision into a form and function that users can and want to use. You never know if your product works until users use it. At the same time, you can't test users using a product and then start to work on a new version of it without analyzing what happened, looking at the things users did and said, and then turning them into actionable recommendations for next steps to make the product better. A report forces you the designer, or your design team, to put all data out there about the product, even if it is just a conversation that takes place among you and your team.

Scientists take notes on experiments that they are often the only audience for simply because this method of reporting on the results of experimentation serves more than just to document the process. The scientist needs to tell the truth about what happened; describe the results in detail; and write up problems, mistakes, and new things learned—anything of relevance to the experiment is documented to help improve future experimentation or to report on the results of experimentation that may become useful knowledge for other scientists.

You have to report for the same reasons. This reporting can be informal, such as a briefing among members of the team in which you get out a whiteboard, list all of the findings from testing the design, and then determine what needs to be done to improve the design. More formal testing, perhaps when other stakeholders (such as managers and directors) are involved beyond the team, should involve a more formal report. This is especially true if others are responsible for giving your team more money to continue to design, or are, in fact, responsible for making final decisions about changes to prototypes.

Regardless of the report's format or audience, a good report answers questions or speaks to the goals you have for your product. It also provides recommendations for what to do if the product, after testing, needs improvement. Such recommendations represent arguments you need to make to convince others—those on your team, stakeholders in your organization, your client—that they should support what you want to do to improve the product.

At the core of any argument, derived from Aristotle's (1959) writings on rhetoric, are three fundamental concepts that must be included for the argument to be successful:

- Ethos: Your argument, including you or your team as the persuader, has to be credible, which is achieved through careful analysis with strong evidence to support your recommendations.

- Pathos: Your argument has to appeal to emotion, which convinces your audience that they need to accept your recommendations because it will make them and their users happy; sometimes you can use negative user experiences to disappoint the audience and make them feel an urgency to accept recommendations (for example, if the audience doesn't accept your recommendations, the users will not use the product or be unhappy).
- Logos: Your argument needs to be supported by facts, based on the *see-say-do triangle*, which provides proof of the problems you've found and also supports your recommendations for fixing those problems.

When preparing your report, you must consider how you will argue for your audience to accept your findings, including your recommendations. It is extraordinarily challenging. We can recite for you multiple instances of when we have prepared thorough reports with sound recommendations based on best practices in design as well as evidence collected from direct user feedback, only to have most if not all of these reports' recommendations not accepted. There are myriad forces at work, ranging from budget concerns, personal opinions of stakeholders or clients trusted more than user evidence, insufficient support for changes, too little time to implement modifications before already set deadlines...the list goes on.

Some stakeholders, for example, may be opposed to your efforts because they don't see the value in UCD. According to Holtzblatt, Wendell, and Wood (2005), "For people who think that customer-centered design is a passing fad, resistance can disguise itself in rational arguments about statistics and time" (294). They will say there is nothing you are doing that contributes meaningfully to successful product development that can't be done by other approaches that are more proven, such as a technology-centered, waterfall design. Ironically, therefore, it is possible, despite your best efforts, to design a product to meet user needs and wants, to work effectively where they will use it, and to then test out a prototype of that product and gather up real user feedback and your audience still be reluctant to "drink the Kool-Aid." Holtzblatt, Wendell, and Wood suggest including doubters on your team, opening up the design process to their contributions, even creating a team design room that encourages transparency and participation (293). We agree that this is a smart, proactive approach, but we also assert the importance again of a rhetorically strong report.

In 2009, during the height of the surge in the wars in Iraq and Afghanistan, a team of developers was tasked with fixing a problem on a military website, the Casualty Mortuary Affairs Operations Center (CMAOC). The website was responsible for providing assistance to the families of armed forces personnel who had been killed in action. The existing site was poorly designed, frustrating its bereaved users who were directed there for assistance. It included hundreds of pages of instructions on how to apply for benefits, confusing navigation, and inappropriate images of coffins for selection, which clashed against tone deaf statements declaring that the military, on behalf of families, was "taking care of business."

There were complaints and calls for change from the public, the developer's organization, and its military client. Developers addressed all this, following the RABBIT approach to analyzing the existing site and its primary users, then constructing and testing an alternative prototype that generated data to identify requirements for inclusion in another prototype.

But they faced a challenge. All their work had led them to believe they created a better website to meet the needs and wants of its intended users, but they had to convince multiple audiences to believe this too. This audience included senior military officials who are often skeptical of change, especially without clear, strongly supported reasons. To make the case, the design team generated a report that took full advantage of the concepts of ethos, pathos, and logos. To see this in action, we've included examples from the report in Figures 9.1 through 9.3. In Figure 9.1's four slides, all taken from the report (read left to right, top to bottom), the developers explain their methodology, providing details on how UCD testing works, supplemented with information about how they conducted their testing of the product, including the user scenarios they created. In the final slide (bottom right), they begin to highlight testing results, focusing on navigational problems users encountered while carrying out the test scenarios.

As the report continues, further evidence about user frustration is presented, accentuated by video highlights, as shown in Figure 9.2.

The evidence presented, showing the major issues users encountered, is not overwhelming but is clear and to the point. It touches on key parts of the site, such as navigation, layout, and content and also involves both performative data (time on task, average mouse clicks) as well as user verbal and observable data (key quotations along with video showing frustration), which helps build a credibility (ethos) that enables developers to make logical (logos) recommendations derived from reasoned, comprehensive testing (bottom right slide).

But the developers don't stop there. They carried out testing to discover an effective UCD solution. That solution, seen in Figure 9.3, is a prototype that enacts the recommendations made prior to it in the presentation. Coming at the end, it serves as an effective rhetorical denouement (pathos), a cleaner, better designed, emotionally appealing prototype built off previously presented evidence in the report. This gives the audience a positive and thus persuasively convincing image of what they will get if they agree to adopt the recommendations.

Of course, crafting a rhetorically strong report isn't always a guaranteed solution. A few years back, we worked with a client to design an online course management system. We conducted significant user research, and in the process of creating and testing multiple prototypes of the system, we recruited and collected feedback from more than 100 users, including college instructors who would use the system to manage their courses as well as students who would go online to read materials, take quizzes, submit assignments, and check grades. The process lasted more than a year and cost our clients hundreds of thousands of dollars; additionally, we committed easily thousands of hours of time to the effort. It was a struggle, but, by the end of the work, we thought we helped contribute to a new system that would offer flexibility to instructors, and thus, at the same time, enormous profitability for the client. The client could cater to instructors from different disciplines that also had different material needs.

At the very end, just when the report was to be reviewed by key leaders for final approval before the product would go from a hi-fi prototype to a fully functioning

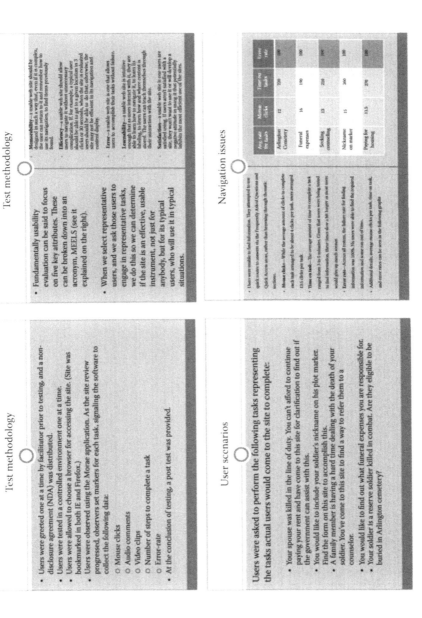

FIGURE 9.1 Report of user testing methodology of CMAOC website. (Courtesy of Dr. Brian Still.)

Usability issue: User frustration

Content and design issues

Contributing to increased time on task and error rates, as well as the frustration that users expressed, were problems with how content was displayed. Users noted (as many video support) the following:

- The long PDFs were daunting and most users did not take the time to read them
- Users were looking to find quick answers via the FAQ or Quick Access areas and quickly became frustrated when they did not find what they were looking for quickly
- The search is AKO enabled and not available to users without an account
- Acronyms were confusing to most users, as they searched for information, they wondered if pages with acronyms in their titles would have what they were looking for
- The site is so vast and poorly organized, users were unsure of what they'd seen before and what was new to them
- Different styles across the site make the users confused about whether or not they were on the same site
- The writing style of the pages was not in a common tongue, users noted that they didn't really understand what it meant

Recommendations

- During paper prototyping of the new design, users confirmed that it is better to see less on the homepage and instead have clear cut paths based on audience type. The least number of choices a user has, the easier it is for them to find the information they seek. On the sections, define clear separate categories where information would reside, and also include a quick access section for frequently referenced documents.
- The site requires a content rewriting as much of the content is written in a style not easily understood by an audience that does not have a military background. Copy is heavy in acronyms and what the users called "legalease." They became very frustrated and felt the site was not compassionate and did not cater to them.
- A search function should be included. Some users stated that would be the first place they would go, but that option was not available for them.
- A clean consistent interface is needed across the website. Each page should include navigational cues, consistent topographic styles, and an intuitive navigational structure. An informational architecture analysis of current content is needed to provide a path toward reorganization of the site.

Content issue: No specific information

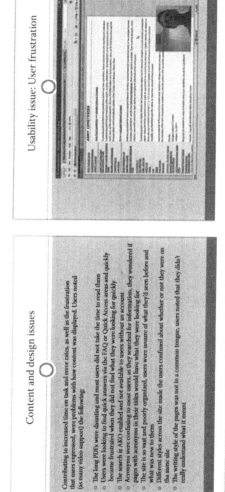

FIGURE 9.2   User frustration and recommendations for action for CMAOC website. (Courtesy of Dr. Brian Still.)

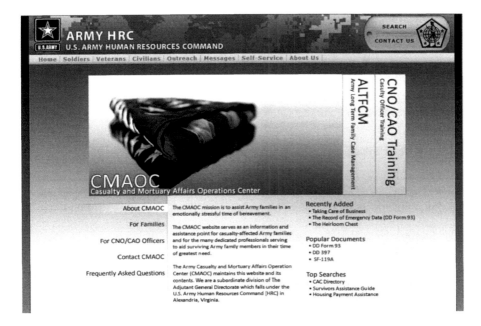

FIGURE 9.3   Prototype for CMAOC website derived from UCD. (Courtesy of Dr. Brian Still.)

product, the client's project manager suddenly died. We had, working with him, created a thorough report—hundreds of pages documenting why the prototype that had been developed would be successful for the client. It turned out that, unfortunately, the project manager was so controlling that not many other people on the client's side had been involved in the system's creation. As a result, nobody in the organization was capable of taking over after the project manager's passing. Because the work had been partitioned off as a special project, it apparently was also controversial internally. Some key stakeholders didn't like it and didn't want it to change the organization's core processes for managing and selling course materials. When the project manager died, so did any impetus to continue the project.

Obviously, you cannot necessarily anticipate something like this occurring, and any report you create would likely not ever have a sufficient mix of ethos, pathos, and logos to make a difference in this case. But this example more than anything else should make clear to you that just as the product you create should always be something you know will be used by real people in real situations, the report you create about that product, its users, and the things needed to make it better for them is also something that will be used by real people in real situations.

Don't think of a report as something you write at the end of a process that just ties up all the loose ends. When you report, design for your audience just as you design for your product's audience. Consider these questions when crafting your argument:

- Who will read and use this report? (Are there decision makers in the audience who don't understand or appreciate UCD?)

- What do they need to know, and what proof do you need to provide for the audience to accept your recommendations?
- Will you prepare a formal report, or can you create something that is informal to get results? Which is most appropriate for the audience?

Formal or informal, short or long, prepare a report knowing that the product you've worked so hard on could succeed or fail because of it.

9.1.1 Report Format

Regardless of the type of report you prepare, your report should be usable, useful, and professional. Include the following key elements in more formal reports to audiences that include people beyond your team:

1. Summarize key findings and recommendations
2. Make the purpose of the test clear (goals)
3. Indicate
   a. History of product development
   b. Where in the process this test takes place
4. Summarize user profiles
   a. Information on users who tested (draw on pretest questionnaire data for this)
5. Explain, if necessary, what UCD is and how it works
6. List test tasks and user scenarios
7. Detail methodology used to gather data
8. Provide results
   a. Results broken down into key issues
   b. Supportive graphs, tables, quotes, and video/audio clips, if available
9. Make recommendations

Should you make a more informal, in-house report (shared just among team members), some of the preceding elements, such as an explanation of what UCD is or background on the history of the product's development, can be excluded. But even in informal reports, it never hurts to adhere to a consistent report format that guarantees what you present is comprehensive and compelling.

9.1.2 Types of Reporting

9.1.2.1 Informal Deliverable to Test Team

If you test early in the process, such as a paper prototype, if the report is just for your team's own consumption, or if you need to move quickly because, for example, you're testing out an MVP so that you can get fast feedback into the design process without slowing down the pace, opt for a more informal report. This can be a stand-up, similar to what is done in agile software development.

The stand-up is often a daily occurrence when agile developers, as part of a scrum or team working on a software development project in an agile environment, provide an update about progress. The agile approach emphasizes teamwork in software development that leads to multiple, fast iterations of software; therefore, lengthy, formal reporting of work isn't possible. The stand-up enables team members to report on their work and get feedback quickly. It is also an oral report.

Desiree Sy (2007) and her team experimented with converting the agile software stand-up into a user testing report. It is a framework you can model for UCD. Because the purpose of reporting on user testing results is to "instigate action," according to Sy, "written text [of a typical report] was not the best medium to achieve this. We needed to convey it to the team in a more timely, vivid, and specific manner" (126). The solution for Sy's team, what we will call an agile UCD report, is to report verbally, "supplemented where necessary by a demonstration of the prototypes or working versions to represent the observed interactions" (127). The material in this agile UCD report is presented verbally in the stand-up with an archive of the verbal presentation transferred into "issue cards"—one issue per card. These cards can be referenced and used by the team as warranted. As Sy notes, "Sometimes an issue card is moved immediately into the cycle planning board, and becomes a feature or design card...The remaining issue cards are tracked on a User Experience board in the same public space" (127) for developers to access as needed. The result, therefore, is a fast, informal reporting style that hits on the key successes, problems, and recommendations learned from testing in a verbal report. The team communicates about the testing of the product, decides on a course of action, and then moves back into development of another prototype. Note cards with issues, such as changes to the product, which the team has prioritized as needed, are moved into the product design cycle for immediate implementation. Other cards are saved for future reference.

You also can move even faster and operate even more informally by using a whiteboard to list key findings from your testing and then working together as a team to determine conclusions and recommendations. If you are on a small team, or even working alone, this may be an approach you can adopt. It may not be effective for larger projects because a whiteboard doesn't offer sufficient space to capture all aspects of a bigger product's design. Groupthink can also be a problem, as it is possible, even in a small group, for one voice, such as a team leader, to dominate and, therefore, tunnel conversation toward conclusions about the data he or she wants. Still, if you can set up a few ground rules to handle this (voting on decisions, for example), a whiteboard report may work for you.

Whichever way you report, even if informally, make sure you know what you learned and what it means; then agree on findings and recommendations. Put these into a document or other storage device so that you have a source of common knowledge everyone can refer to when working on making changes to the product. Sometimes at EyeGuide, when developers are moving quickly, taking a photo of the report just authored on the whiteboard, as Figure 9.4 shows, is enough of an artifact to use for later reference.

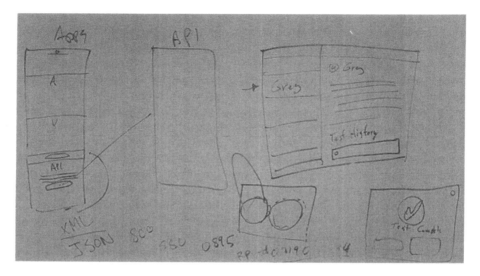

FIGURE 9.4 EyeGuide whiteboard notes after user testing of focus prototype. (Courtesy of EyeGuide.)

9.1.2.2 Presentation Format

If your report needs to be delivered to an audience outside of you or your team, such as managers, designers, or other stakeholders, including clients, we find that a presentation is a useful deliverable. It functions well to accentuate key findings, providing a brief but comprehensive overview of the methodology you used, offering also evidence for support—graphs, data tables, even audio/video highlights—without being so burdensome in terms of length or amount of material that the audience won't want to read it.

Presentations are valuable for several reasons:

- They hit the key, high notes of your findings.
- Video and audio files can be embedded into them.
- They offer a useful blend of detail but also efficiency—just enough but not too much information, allowing for easy sharing among different groups in an organization.

You have to report something about what you've learned from testing your product with users to someone. Whichever format you choose, use the report as an opportunity not just to document what you learned, but also to make recommendations for improving the product for another round of development and testing. Box 9.1 lists important UCD report best practices to help you develop your report approach.

BOX 9.1   UCD REPORT TIPS

1. You're making the case for issues that need attention—the report is a persuasive act.
2. Including video and audio clips of users in action using the product can be especially effective. Carter and Yeats (2005) note in their study of client reaction that when video and audio clips of users were shown, clients were more likely to be persuaded by the recommendations made in testing results reports. The video and audio of the user is ethos, pathos, and logos neatly packaged together. The user, as consumer, has credibility, the video and audio shows the experience and that is compelling evidence, and the user satisfaction or displeasure, the emotion, is plain to see.
3. Combine qualitative and quantitative data and allow each to strengthen the other. Quantitative data builds your logical argument—gives "hard facts" and numbers—and qualitative data builds your emotional argument—gives the client an opportunity to see (or hear) users dealing with the issues in the interface or product.
4. You can't identify 40,000 problems and expect your audience to implement all of them; in fact, consider just how many or which ones they're likely to implement, and consider constructing a report that you emphasize with findings for those problems that require immediate attention or are a big deal.
5. Don't forget the good things—your job isn't just to show problems. It is to let the real users, performing real tasks in real scenarios, reveal by their actions, their words, your observations, and then analysis what is going on and what the issues are.
6. Don't write or present a book—in other words, shorter is better; of course, you shouldn't skimp on substance, but your audience more than likely won't implement every one of your recommendations, and they definitely won't read every line of a 35-page report or listen attentively to a 60-minute presentation.
7. Presentation slides should be able to stand alone, that is, without the need for narration. Each slide should be clear, be effective, and contain complete and focused information. Visuals—images, video clips, charts—will really help the presentation "pop" and give the client different ways of seeing the data, the analysis, and user feedback.
8. Keep in mind that some in your audience, especially those from outside your UCD team, will likely be wary of the validity of the results when you have such a small n (test group/number of users). Remind them that the law of diminishing returns plays a big role in determining this number, and there is an equation to determine the ideal size of n to test. Here it is also useful to explain what UCD is and how small batches of users, interacting with multiple iterations, play an

important role in designing the product, not validating statistically the product's likely acceptance, for example. There will be a time and place for that type of testing with a larger user population. What you're doing is designing a product for users, and you are reporting on a select group of users' feedback on a prototype to help you understand what's right and wrong so that you can iterate a new, better version.

9.2 ITERATE

In Chapter 3, we discussed Morgan and Borns' (2004) concept of UX and thus UCD as a 360 degrees, circular, not linear, activity. At eBay, the process of iterating changes to the company's main website is an ongoing, never-ending affair. This is true for other organizations as well. Whether you are building a new product, revising a product, or maintaining a product to keep it effective for its intended users, you must always evaluate user feedback, account for changes in the situation, and, accordingly, find ways to test and implement necessary, new versions of the product.

This doesn't mean throwing out a product and starting over. There can be drawbacks to this because the product's audience may have gotten used to the previous version, despite its flaws. You don't want to mess with mental models. Rather, think about iteration, for a new or existing product, as a mandate to monitor the product's use all of the time. Set up a process whereby you constantly collect user data that forces you to experiment with new ways to improve their experiences.

Iterate. Iterate. Iterate. Even if your budget and time are limited, find ways to iterate the design as much as you can. This increases the accuracy of your design decisions based on user feedback, and it guarantees that users remain the focus of the entire process. Anything in which users engage with the design is a test. Test early, and test often. Testing also ensures that users aren't just giving advice or opinions. We want users to be the focus of the design process but to act as users. Testing the design by putting users in realistic scenarios gives comprehensive knowledge we can rely on, which is what we see, what users think and say, and what users do.

Let's return to EyeGuide Focus again for our example of iteration. You can see throughout this book the many times we've referenced the prototypes EyeGuide created, tested, and then iterated the process again. And those are just the ones we referenced. All told, EyeGuide iterated dozens of iterations of Focus software as well as hardware and even support materials. To this day, Focus undergoes evaluation and improvement through iteration. The wireless setup for connecting the headset to the tablet, after testing and feedback, was recently converted to a wired setup. This occurred after Focus was already on the market and being sold and used. Because users asked for it, and because EyeGuide developers learned that such a wired approach was cheaper and more reliable, Focus was iterated to feature that solution.

User feedback is at the heart of UCD, and to make it constant and productive is to assume and then require constant iteration.

9.3 DEPLOY AND EVALUATE

The universe never ends, and we discover new things about it almost every time we look at it. So it should also be for products. We want to evolve our knowledge of the universe that users inhabit, down to the micro-level of where they actually live and work, and we want to make our products better for users to use. UCD is about evolution, about growing and building upon our knowledge of users, of products, and of the methods and processes we use to build UCD products. Deploy the product, but don't stop learning about it, users, the environment, and the things you can and need to do to make products better for their users.

There is an enormous return on investment (ROI) in evolving products continuously to meet user needs and wants. Data about their preferences, about how they actually use things, what they can't do, what they can do—all of this and more becomes knowledge that we use to improve products and our understanding of users, which can be profitable not just economically, but also technologically and culturally. We're supposed to make things better, supposed to advance. UCD doesn't just make products better. It makes users better, and that makes their world and ours better. Research, assess, balance, build, test, report, and then do it again.

TAKEAWAYS

1. UCD is a recursive process that lends itself to proactive monitoring and ongoing improvement of a product to meet user needs and wants.
2. RIDE (report, iterate, deploy, evaluate) helps inform you, your team, or others about your UCD projects and enables constructive, iterative modification and evaluation of your product.
3. A good UCD report answers questions or speaks to the goals you have for your product. It also provides recommendations for product improvement. Importantly, a good UCD report uses a combination of ethos (credibility), pathos (emotional appeal), and logos (logical appeal) argumentation strategies.
4. Don't think of a UCD report as something you write at the end of the process. Design the report for your audience just as you design your product for your user's audience. Know that your product could succeed or fail because of your UCD report.
5. There is often organizational pushback or other exigencies that force modifications to—or even results in failure to adopt—UCD recommendations: monetary, political, material, and functional constraints.
6. Including video and audio clips of users using your product can be an effective report strategy.
7. Iteration should be thought of as a mandate to monitor the product's use all of the time. Set up a process whereby you constantly collect user data that forces you to experiment with new ways to improve their experiences.
8. UCD doesn't just make products better. It makes users better, and that makes their world and ours better. Research, assess, balance, build, test, report, and then do it again.

DISCUSSION QUESTIONS

1. Imagine that you've gone through the UCD process for a client and are preparing a report of your findings and recommendations. What types of information would support ethos (credibility), pathos (emotional appeal), and logos (logical appeal) arguments?
2. What are potential monetary, political, material, or functional constraints that might lead to alteration or nonadoption of UCD recommendations? What might you do to plan for or mitigate these in advance?
3. How would you choose which UCD problems and recommendations to include in a UCD report? (You can't share them all.)
4. How would you explain the validity of your UCD results if you had a small *n* (test group/number of users)?
5. Why should you constantly iterate your design? When do you know whether it's time to improve on a current design or create a whole new product to address evolving user needs?

REFERENCES

Aristotle. *Ars Rhetorica.* Edited by W. Ross. Oxford: Oxford University Press, 1959.

Carter, J., and D. Yeats. "The Role of the Highlights Video in Usability Testing: Rhetorical and Generic Expectations." *Technical Communication* 52, no. 2 (2005): 156–162.

Holtzblatt, K., B. Wendell, and S. Wood. *Rapid Contextual Design: A How-To Guide to Key Techniques for User-Centered Design.* San Francisco: Morgan Kaufmann, 2005.

Morgan, M., and L. Borns. "360 Degrees of Usability." Proceedings of the Conference on Human Factors in Computing Systems, Vienna, Austria, April 24–29, 2004: 795–809.

Sy, D. "Adapting Usability Investigations for Agile User-Centered Design." *Journal of Usability Studies* 2, no. 3 (2007): 112–132.

10 International UCD

Groupon, a web service that buys products and deals at group rates and sells them for a lower price to local customers, began in 2008 under the name ThePoint.com and quickly became an international operation. By 2011, Groupon had successful operations in many European countries and was continuing to grow. However, Groupon had problems in China—problems that have become a textbook case of mismanagement. But, more than mismanagement, one of the problems Groupon faced was not considering the local culture, which resulted in failed relationships with its various users. Zhu (2011) explains that "Groupon's aggressive sales tactics" insulted local sales persons who "told the company's sales people to calm down and come back later with more realistic expectations" (par. 8). Further, "Groupon insisted on using mass email marketing, despite being warned that Chinese people seldom read that type of email" (par. 8). Creating usable products, as we've noted before, takes more than wanting users to like them. It means conscious, reflective, and iterative development—even for a successful business trying to stake a claim on a new country's user base.

Considering international users is imperative to good UCD. Our world and work expands boundaries in ways it never has before, and it will continue growing and morphing. Many intercultural design theories provide "how-to" lists for international design, but this chapter does not do that. Although intercultural communication design theories are excellent contextual starting points to begin conceptualizing products, nothing can replace user-specific research. There are no quick and dirty ways to do UCD, let alone international UCD. "How-to" lists would be counterproductive to following UCD principles. As in all UCD projects, the planning stage of design is the most important and strategically necessary step.

In this chapter, you will learn how to adapt, when necessary, the RABBIT method to design for international users. Specifically, we discuss how to research international users, understand the scope and context for your design, design MVPs and prototypes, and test with international users.

10.1 WHAT IS INTERNATIONAL UCD?

Nielsen (2011) describes international usability as "the effectiveness of user interfaces when used in *any other country* than the one in which they were designed" (par. 2). International UCD can be thought of similarly. In the context of international UCD, the people or organization designing the product are not members of the same country, region, or, perhaps, language group as the product's potential users. We see many examples of big corporations producing hardware, software, or communication for international users. Take, for example, Apple. The iPhone is not just a U.S. phenomenon, but also an international phenomenon. Users all over the world flock to the Apple store for the newest model. Therefore, Apple must be concerned with not only how users in the United States use the product, but also how users in Brazil, Japan, or South Africa will use the product.

As a designer of international products, you need to appeal to larger user groups and consider how international users will use your products. Although the process of researching and developing products may seem the same as any UCD process, more specific analyses of the project and users are needed because you need to be familiar with the culture to design for it. As such, you need to examine the project's scope, context of use, and resources available. Scope refers to the range of user groups that the product is meant to reach. Context of use is how and under what conditions users will engage with the product. Understanding resources is twofold: We need to consider not only what resources are available to users (which is connected to the context of use), but also what resources are available to you while designing the product. Only once you and your design team understand these factors can you begin designing the product. The following sections discuss how to understand the role of scope, context, and resources for international UCD using the RABBIT method discussed in this book.

10.2   RESEARCHING INTERNATIONAL USERS

The differences between researching users of products in your own country versus international users are not much different. As emphasized in Chapter 4, the only way to learn about user needs is to work with users directly: Observe what they do, listen to what they have to say, and analyze their performance of tasks. The major difference with researching international users is that access to the target user population and resources for conducting research may be limited. However, shortcuts in international user research are not a good idea as we learned from Groupon's China experience.

Although you could study intercultural communication theories to understand general cultural differences, they have their own limitations. Two of the most cited intercultural theorists used in communication are Edward T. Hall (1976) and Geert Hofstede (2010). Hall developed the theories of high-context and low-context cultures, and Hofstede breaks down cultures into five cultural dimensions to describe national cultures. Both of these theories can provide general considerations for cultural analyses, but they don't tell you enough about who a specific user population is and what its needs are. Further, these theories can lead to stereotypes that get in the way of creating user-centered products.

User experience experts agree that nothing replaces field research. Rourke (2013) explains that "there will always be a need for an ethnographic approach to see how people currently solve the problems that your solution hopes to address" (par. 7). Quesenbery and Szuc (2012) note that to understand culture, "we have to look beyond individuals to see the whole ecosystem around them" (35). Ethnographic research typically involves spending an extended amount of time embedded in a culture to observe how it works. This, however, is rarely practical for UCD projects unless you are well funded and have a very understanding boss. A better way to think about this type of research for UCD is as quasi-ethnographic fieldwork. Instead of spending months observing a culture, you may get a week. In this time, you want to learn everything you can about users, including their living and working environments. Elements you may not have to worry about in your home country may greatly impact international users; these include economic and technological infrastructures, education and technology literacies, and devices available for technology use. You may be able to get a general sense of how these factors play a role by looking at Internet usage reports and other secondary sources, but seeing how people work within the confines of their culture's contexts takes fieldwork.

What do you do if fieldwork is not an option? Sometimes, fieldwork is not feasible due to lack of resources (both time and money). In this case, consider consulting UCD researchers already in the field. In *Global UX: Design and Research in a Connected World*, Quesenbery and Szuc (2012) discuss the benefit of working with local partners or researchers and consultants who are currently in the country or culture you need to research. Not only do local partners eliminate the need to travel to your international users' locale, but they may also have knowledge and expertise about the culture that can help you understand user needs, questions to ask users, local constraints for product development, and even ways to recruit users for research (and test later in the design process) (Quesenbery and Szuc 2012). These local partners can do the field research for you and, perhaps more importantly, they can connect you to users you would not otherwise have access to.

Remote research is another alternative to traveling to your users' locale. Remote research uses technology to connect to users. For instance, if you want to interview international users but cannot conduct interviews in person, you may opt to use video conferencing tools, such as Skype, GoToMeeting, or even FaceTime, to speak with users one-on-one. Further, using digital diaries (discussed in Chapter 4) can provide information about how users go through their days with or without your product so you can collect data before you begin revising your design. Digital diary tools can also prompt users to take photos of their work or use environments, thereby providing you with visual data as well.

You have one last option for working with users if you cannot travel. Rourke (2013) suggests finding users in your home country that are representative users of your international target audience. For instance, if you live in a large city, such as New York or Los Angeles, you may be able to find users of different nationalities who speak the culture's language and are familiar with the customs of the international community you are designing for. There are a few factors to keep in mind though. First, although someone's country of origin may be part of your target audience, they may have assimilated in ways that would change the way they use your product (especially digital technologies). However, if you are looking for linguistic insight, this may not matter. Understand what type of feedback you are looking for, and understand the limitations of local users. Second, make sure the users you are researching are truly users of your product. We know that culture is more than language and nationality, and our users may be part of more specific cultures such as workplace cultures. The local users you select should resemble these target user culture(s) as closely as possible. Finally, although these users could provide great insight, you should still get feedback about your product from your international users. For instance, if you want users to interact with prototypes, local users may be a better choice if the prototype is not digital or cannot be shipped easily. Rourke

(2013) also suggests using local users for pilot testing before beginning more global research.

In addition to field research, you should consider what other types of data you have access to. Putnam et al. (2009) provide a methodology for creating personas and scenarios based on existing research. These designers were working on a mobile phone directory, MoSoSo, for Kyrgyzstan, a country with rapidly growing mobile device use. Because it lacked the resources to do its own research, the design team used existing survey, focus group, and interview data providing census and mobile use information not focused on a specific process or design. Through statistical and qualitative analysis, Putnam et al. (2009) found that the data provided insight into user characteristics and motives, which helped them develop three distinct personas and different scenarios for using the mobile directory that assisted designers in conceptualizing the directory and designing prototypes for testing.

As with all cases of user research, you should triangulate your research so that you can see what users are doing, listen to what they have to say about their use and experience, and watch them perform tasks (the *see-say-do triangle*).

10.3   UNDERSTANDING INTERNATIONAL USE SITUATIONS

Like researching users, assessing the situation for international UCD follows the same principles of analysis as designing for domestic users. You should analyze the following:

- Product function
- Use environment and potential constraints of that environment
- Organizational needs (if there is an organization you are working with)
- Product competitors
- Materials to design the project
- Product content

For international UCD, we want to examine similar questions about the contexts listed above as in Chapter 5, but we also need to acknowledge different conditions we don't think about when working on domestic projects.

First, when thinking about functional needs for the product, you should look at how users plan to use the product or how they are currently using an existing product that your design would replace. This may be more difficult for an international context due to limited resources, but even when resources for fieldwork are unavailable, existing research can be valuable to understand users. As discussed earlier, researchers analyzed user priorities and needs using existing research about mobile phone usage in Kyrgyzstan. One thing that Putnam et al. (2009) understood was that mobile users were adopting mobile devices so they could make calls anytime during the day and from any location, which landline telecommunication could not consistently do. The second thing they understood was that the mobile device and the MoSoSo directory needed to fit with a social network framework as "face-to-face social networks are critically important in the region...[and] serve as an avenue for gathering and sharing information, assistance, and goods" (Putnam et al. 2009, 54). Thus, the directory design needed to function as not only a number directory, but also a way to establish and maintain the social networks valued in Kyrgyz culture.

In addition to function, environmental factors play a crucial role when designing for international users. Chapter 5 outlines important questions to consider for environmental analysis. One question that is missing but particularly important for developing areas is, does the infrastructure of the users' locale support the design? Sticking with the MoSoSo directory example, we could ask, why not create a website instead of a mobile directory? The answer is based on environmental factors, such as the telecommunication landline infrastructure not being able to support rural user needs, more people switching to mobile devices for better telephone access, and fewer people adopting or accessing the Internet with or without computers (Putnam et al. 2009). Thus, a website would not fulfill user needs based on the infrastructure available. Wireless mobile devices, however, provided access where the landline infrastructure failed. Further, incorporating the directory interface into the phone would be the most convenient location for this resource. Kyrgyzstan is not the only example of rural users adopting mobile devices. This is a trend among developing nations. When creating content for the web, for instance, understanding the context and infrastructure for using and accessing information for international users is critical.

10.4 CHOOSING BETWEEN INTERNATIONALIZATION AND LOCALIZATION

Now that you have our research and assessment data from RABBIT, you need to decide how to incorporate users' needs and wants with what you would include to create a usable product. As discussed in Chapter 6, just because users indicate they want the moon and stars doesn't mean the moon and stars make a product usable for a given task. This is when we go back to our user research and contextual analysis to pinpoint the need-to-have versus want-to-have product features.

For international UCD, one of the balancing factors for design is considering the scope of the design. Scope can be thought of as the level of diversity within the user population and, therefore, how focused the target user base is for a product. Scope can be broadly broken down into internationalized and localized groups; however, within these levels, a more focused scope of users can be found. For instance, a localized design can be regional and limited to a continent, or it can refer to a cluster of countries that have similar cultural needs. A localized scope can also refer to a country or a more limited community within a specific country. How broadly or narrowly you define the scope of the project depends on the goals of the project and its expected representative user population.

10.4.1 INTERNATIONALIZATION

An internationalized scope refers to creating a product that is appropriate for all users, despite their countries or origins. Why would an organization choose to develop a product that is less specific to one of its user groups? If you are an organization with a large user group, it may not be feasible to create documentation or a website for each user's country of origin. However, one set of instructions or a website made accessible to an international group would be viable. Many government websites, for instance, use the same webpage layout with the ability to translate, switch languages, and search the website in different languages.

Even International Organization of Standardization (ISO) symbols, which are symbols designed to have the same meaning across cultures, are used in internationalized websites and products. Images are often used in place of text for internationalized products. Ikea is a good and popular example of using images to internationalize documentation. Ikea assembly instructions only use pictures and diagrams. This way, despite their origins, users can follow the same instructions.

Using ISO symbols are also important for travelers. To navigate a new city or country, usable symbols are needed (Figure 10.1). Singapore, for instance, uses several warning and informational graphics around the country. Although Singapore's official language is English, people from all over the world work, live, and travel throughout the city-state. Therefore, signage that can be understood across cultures and languages is imperative.

This type of communication design is beneficial not only for users in different countries, but also for users in the same country who do not speak the dominant language or languages. The benefit of internationalization, then, is to create one product to serve the needs of many, varied users while saving organizational resources.

Some experts go so far as stating that all websites should be created with the understanding that international users will visit the site. Bevan (2001, 2009) suggests using ISO standards for HCI and usability because products and information are more likely than ever before to cross geographic and cultural borders. St. Amant (2003) stated that the growth of international Internet users means communication

FIGURE 10.1 "No walking" sign at Singapore's National Orchid Garden. (Courtesy of Dr. Kate Crane.)

designers need to create "design strategies for effective online materials" for international users, including creating communication with the expectation that content will be translated, that users will have different bandwidth speeds, and that many users need to print materials (15). Thus, even when you think you are designing for a domestic audience, international users could still potentially use your product (especially in the case of websites or other online material that is easily accessible across borders); therefore, using an internationalization approach to building your design is good idea.

10.4.2 LOCALIZATION

A localized scope narrows the product design to focus on a more specific user group. Designers of instructions may do this by writing instructions in two or more popular world languages, such as English, Chinese, and French. However, some designers may be even more specific and recreate entire websites rather than try to accommodate different user groups by merely changing the language. McDonald's Corporation is an excellent example of a company that localizes websites, and it updates these localized sites frequently.

In November 2015, the McDonald's U.S. site included an American football background and theme. During the fall season in the United States, one of Americans' favorite pastimes is watching football. In fact, during the U.S. holiday season, between late November to early January, major national holidays feature college and professional football games on major television stations. Thus, McDonald's teamed up with the National Football League (NFL) to offer a promotion for eating at McDonald's. The main homepage image featured a still image (which turned into a video when selected) of Jerry Rice, a famous football player who played in the NFL from 1985 to 2004. Jerry Rice is recognizable to American football fans, but international visitors to this website may not have recognized this cultural figure.

In February 2016, the website infrastructure was the same, but it promoted McDonald's new all-day breakfast menu. The background was changed from the football pattern to black. Like the football promotion, a video (a static image until selected) was the main feature of the homepage. Two of the three bottom links advertised the same things as the November site: a mobile app and a mailing list.

The U.S. website was only presented in English. Although the United States does not have an official language, English is the dominant language used in public life. However, although the website was in English, the official rules of the McDonald's football promotion could be accessed in either English or Spanish, the second most commonly used language in the United States.

McDonald's, however, recognized the need for a different design for its international users. The United Arab Emirates (UAE) McDonald's homepage is an example of this design. One of the first differences noted between the UAE and U.S. homepages is that the logo and top menu options for the UAE site were aligned on the right-hand side instead of the left. This different alignment is necessary because Arabic is read from right to left. However, unlike the U.S. site, the UAE site gives users the option to read the page in Arabic or English. When English is selected, the menu switches to the left side of the page (as well as the written content aligned to the left) although image placement remains in the same format. The right-hand side menu only appears when the cursor hovers over the McDonald's logo on the top of the page.

Images in the UAE are also different. As the main image loaded in November 2015, animation was used to introduce the "Spanish hamburger," and cartoon characters moved into the website to surround the hamburger. However, the animation stopped after the introduction and was static while the site was used. The word "Halal" was circled in the left-hand corner near the image header text. This spotlights McDonald's commitment to serving Halal products. Halal refers to meat that has been prepared according to Muslim laws. Halal meat needs to be certified by a governing body that inspects products to make sure they are following the standards for meat butchering. On the UAE website, not only was Halal confirmed, but also current inspection certificates were viewable. The inclusion of Halal information is not only respectful of the culture, but necessary if McDonald's wants Muslim customers in its UAE restaurants.

In February 2016, the theme of the website (like the United States') was different. A promotion spotlighting Neymar Jr. was featured on the UAE homepage. Neymar Jr. is a popular Brazilian professional footballer (Americans know them as soccer players). The promotion advertised a chance to meet the player, get a signed t-shirt, or be in a television commercial. A picture of Neymar Jr. was placed right of the text discussing the promotion. The bottom of the page had three featured links (like in the United States). One was the same as the November 2015 website, which promoted a "build your own burger" offer. The second link was about food quality and led to an image of various food producers and preparers (all men). Each figure could be selected to learn more about its role in McDonald's food production. A short video was also available below these interactive images. Finally, the third link displayed different Big Mac versions with the word "Halal" clearly displayed.

Finally, let's examine McDonald's Mexico website. The November 2015 version opened the page with a lot of animation. The introduction screen was a moving graphic of Angry Birds to advertise the latest McDonald's promotion. When exiting out of the introductory page, animation was still a primary design choice. On the homepage, an image carousel displayed family-, promotion-, and product-oriented images across the center of the page. On the bottom of the page, more animation appeared. A navigation carousel, one that bookmarked the pages in the main carousel, was used. The page's menu was located across the top and right-aligned. There was no option to change the language on the page.

In February 2016, the same carousel was used, this time featuring Hello Kitty and monster trucks as Happy Meal toys. A bright, pastel color palette was used, and many food images were displayed. Clearly, despite a few details in the image carousel, not much changed about the website.

What can we learn from McDonald's? Looking at its U.S., UAE, and Mexico websites, it is clear that designers understood the social, economic, and technology preferences of each culture. The designers went beyond translating pages to a different language; the focus of the communication design was specific to the values and customs of each culture (see Figure 10.2 for a comparison of U.S., UAE, and Mexico websites). In a matter of four months, the websites changed in theme, but the

	Language	Menu orientation/ structure	Homepage theme	Multimedia
U.S.	English (contest rules in Spanish)	F-Pattern menu: left main menu and menu items across the top. Main promotion is features in center of homepage.	America Football (Nov. 2015). All Day Menu promotion (Feb. 2016).	Video and links Football pattern background in Nov. 2015 background. Black background in Feb. 2016.
UAE	Arabic and English	F-pattern menu with right/left menu appearing when cursor hovers over McDonald's logo.	Spanish Burger—new menu item (Nov. 2015). Neymar Jr. promotion (Feb. 2016). Halal clearly labeled in both versions.	Animation to static page. Links to more pages. Video available but not prominent. Red background for both versions.
Mexico	Spanish	Menu across the top of page. Image carousel in the center.	Carousel featuring families, a game promotion, Angry Birds, and menu items (Nov. 2015). Game promotion, happy meal toys, and menu promotion (Feb. 2016).	Animation and image carousel. Multiple colors used.

FIGURE 10.2 McDonald's country website comparison.

layout and cultural markers remained the same. Look at these webpages now as you are reading. What has changed? What has stayed the same? After looking at these images, take a look at McDonald's' other international websites. How do they differ from the three analyzed here?

When choosing whether to internationalize or localize your product, the decision really should come down to who the target users are and how their goals can be met in the design. Sometimes, user goals, beyond cultural considerations, can be addressed with an internationalized design, or enough user goals can be met with an internationalized design to accommodate your or your organization's resources for the project. However, when you have the resources and your product benefits from localized design for effective use, localization is a safer bet to satisfy user needs.

10.5 DESIGNING FOR INTERNATIONALIZATION

The key to designing for internationalization is using the most generally accepted principle of design. Nielsen (1996, 2011) claims that, although there are details that need consideration for international sites, major website features remain usable across cultures. These include following the "F" pattern for organizing website information where readers fixate on the top menu, on the left-hand menu, and through half the middle page (see Chapter 6 for a more detailed discussion). In a study comparing website use in Australia, China, and the UAE, Nielsen (2011) found that, although the "F" pattern may take on a mirror image for languages that read from right to left (such as for those who speak Arabic), these international users still read following an "F" pattern. Nielsen also found that, although language orientation may be different for users in the UAE, all user groups were able to read left-hand orientation websites effectively.

When designing an internationalized product, multiple user languages must often be accommodated on a single website. An effective way to do this is to design content that is easy to translate. Nielsen (2011) notes that one of the differences between English and Arabic is that English is more concise than Arabic, thus more room is needed to accommodate translations into Arabic. Specifically, if "designing application components such as dialog boxes," Nielsen writes, it "is [important] to leave room for labels and hints to grow by at least 50%" (sec. 3, par. 2), so that languages, such as Arabic, requiring more space can still be displayed effectively.

Preparing for translation also means making search and help functions accessible. Specifically, Nielsen (2011) recommends ideally creating a search function that is multilingual. If multilingual design is not an option, allow both American and British English for search functions and "be forgiving of typos" (par. 17).

Finally, when designing for an international audience, be careful of using icons or gestures that could be offensive or misunderstood by a cultural group. Nielsen (1996) suggests avoiding finger-pointing cursors (pointing at people and things in many cultures, such as China, are highly offensive) or icons of feet, not using culture-specific metaphors (such as baseball references), and not using culture-specific icons that may not be used in other cultures, such as a dining room table or mailbox.

10.6   DESIGNING FOR LOCALIZATION

Everyone wants a quick way to learn about a group of users and apply this knowledge to design. It would certainly be easier and faster to develop new products. There are many studies that look at similarities among cultural groups. Although we will discuss a few common features to consider for localization, we will not give you hard and fast rules for designing for any given group of people. Why? Users are different and deserve to be represented appropriately. Falling back on stereotypes to design for users devalues the UCD process.

You should be aware of features that differ between cultures, including the use of color, language and dialect preferences, orientation of text and document layout, culture-specific gestures and hand signs, religious and social codes of conduct, and the literacy of a group (e.g., reading literacy, computer literacy). Barber and Badre (1998) coined the term "culturability," or the coming together of culture and usability, to discuss how cultural markers (such as color, webpage orientation, types of hyperlinks, etc.) turn into usability expectations within a given culture. The UAE McDonald's website's use of orientation changes between Arabic and English web menus is one example of how usability is specific to the language (and therefore culture) of the site's users. Barber and Badre (1998) find that "cultural markers can be cultural and/or genre specific and then be used to implement culturability guidelines," meaning that using cultural markers can help users of specified cultures use products (10). Additional studies, such as Callahan's (2005) comparison of university websites across cultures, Würtz's (2005) comparison of websites from high- and low-context cultures, and Singh and Baack's (2004) comparisons of U.S. and Mexican websites, all correspond with Barber and Badre's assessment that designing for different cultures poses a responsibility to create usable products for your target audiences' culture. Therefore, part of your research and testing should be gathering feedback about these features to determine what elements are expected, preferred, and usable.

10.7   BUILDING AND TESTING PROTOTYPES FOR INTERNATIONAL USERS

Now that you understand how to scope your international project, it is time to build a prototype. Using your user research and situation analysis, you can incorporate what you know about your users (and their cultural preferences) to build agile and paper prototypes that test how well you are addressing users' needs.

10.7.1   MINIMUM VIABLE PRODUCTS (MVPs)

Chapter 7 provides a thorough overview of the MVP, which you can reference again if necessary. We're going to use an MVP to demonstrate how to create an effective, localized product prototype.

Let's use JTC UX Consultants (JTC) as our example company. JTC is interested in expanding its consulting services to India, and it thinks Indian business owners would be interested in its consulting services. However, JTC wants to make sure it can attract these potential customers to its website before putting more resources into not only the design of the website, but also the overall operational costs of opening up a new market in India.

Instead of designing a more extensive prototype and recruiting users to test it, JTC first started with a hypothesis: Business owners in India will be interested in using our services to better understand user needs for their products. To test this hypothesis, JTC created an email blast to Indian business owners to see how many people were interested in its services and would visit the company website (Figure 10.3).

This email is simple. It gives a brief introduction to JTC's services and provides a hyperlink that potential clients can click on to visit its website. This email is a simple measurement of interest in the product and provides a quick test (just a click of a button) to see how many business owners would be interested in the company's website. By looking at website analytics, or even Google Analytics, JTC can see how many visits this early concept website receives from the link provided in the email. It can also decide what the baseline number is for how many website hits indicates enough interest to invest more resources into the design. For instance, would 2000 hits be worth the investment? 5000? JTC can compare the rapid data

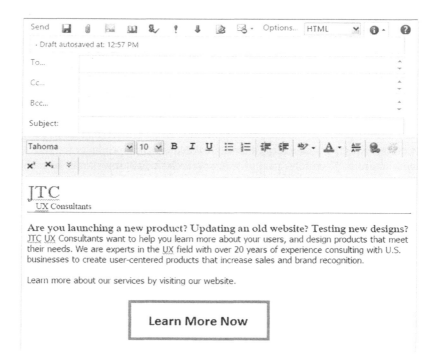

FIGURE 10.3   MVP email blast. (Courtesy of Dr. Kate Crane.)

collection results to this baseline to decide if it should go to the next step: paper prototyping the new website.

10.7.2 PAPER PROTOTYPING

JTC still has more work to do before designing its website, and this step is even more complex because it requires users to engage with the prototype. Paper prototyping, as you remember from Chapter 7, requires creating your design on paper first and asking users to interact with that design. Users can even help create the design using prototyping tools. Using a user profile derived from researching users, JTC then hired an Indian-based consultant to recruit five users in India to perform paper pro-totyping for its new website design. These five users included a 30-year-old female software developer, a 40-year-old male IT consultant, a 25-year-old female computer engineer, a 20-year-old male university student, and a 50-year-old male government web content developer.

JTC created prototype labels to represent items designers thought users would most like to see on the company's website. These labels included menu and link options (i.e., "Home," "Testimonials," and "Contact Us" links as small sticky notes) and labels for larger content items such as pictures and text (Figure 10.4). Users were given the following scenario: *You are a business owner looking for a user experience consultant to help rebrand a word processing cloud product. You know that you*

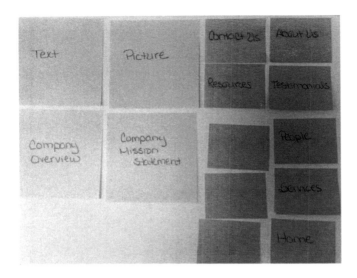

FIGURE 10.4   Paper prototype labels. (Courtesy of Dr. Kate Crane.)

have heavy competition from larger companies, but you believe your product can be competitive and that there is a market for new, better, cloud-based word processing programs. You type UX Consulting companies in a web browser and find a link to JTC UX Consultants in the list of results. You click on the link to its website. Your task is to place the following "links" (i.e., sticky notes) on the blank white sheet of paper to represent where you think menu items, hyperlinks, and content should go. You do not have to use all notes. There are extra blank sticky notes available for you to add additional content you find missing. Users are then given time to design a website that makes sense to them.

What did JTC designers want to learn from this activity in its effort to build an effective website for an Indian market? First, would Indian users expect information to be placed differently than where it would appear on U.S. websites? For instance, in Figure 10.5, the prototype website has a menu across the top of the page. Many U.S. websites will have some links at the top of the page, but it has become popular to have the menu on the left-hand side of the page. Would this expectation also be true for Indian users?

Second, would Indian users have different preferences for the order of menu items as well as the actual subject matter of the menu items? Testimonials, affiliations, contact information, and other information showing the company's trustworthiness may prove beneficial. For instance, Sharma (2015) reports that Indian users found group-buying websites more trustworthy when "multimedia, security certificates/ logos, contact information, and social networking logo" were located on the site (175). Finally, although the prewritten labels did not indicate multiple pictures, testers allowed users to write in information that was missing, such as multiple pictures of the company's employees, clients, and work activity on a rotating carousel (represented by < and > symbols). Kulkarni, Rajeshwarkar, and Dixit (2012) found that "Indian culture is expressed using more pictures than language" (19). These

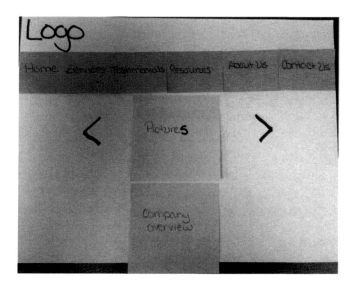

FIGURE 10.5   Results of paper prototype test. (Courtesy of Dr. Kate Crane.)

examples told JTC designers what Indian users would want the company to say about itself to create credibility and also provide necessary content.

10.8 CONDUCTING USABILITY TESTS WITH INTERNATIONAL USERS

You have version 1.0 of your product, and you are ready to test it with your international users. Where do you start? This question is more complicated than you might think because you need to decide where you will test; how you will recruit users; and how culture will affect the development of your scenarios, tasks, and testing methods. As in the case of researching your international users, you will find it easier to work with UX groups, consultants, or agencies to recruit users and understand the contexts of the cultures you are recruiting from. Without this resource, your testing experience may be much more difficult.

10.8.1 LOCATION

You first need to decide where and how testing will take place. Will you travel to the users' locations to test? Will you perform remote testing from your location using screen capture and video technology? Each approach has costs and benefits, and in either case, you will need help to recruit users and set up the test environment.

Barnum (2011) explains that testing in users' environments can give evaluators insight into environmental contexts (such as when testing in users' homes) and help avoid technology issues that may arise. However, traveling to users' locales and finding spaces to test can be expensive. Remote testing, on the other hand, is less costly, and you may be able to test more international users because you will not travel to individual locations to meet users. But, you may not be able to observe user behavior as thoroughly (because you will likely only collect audio and screen capture data) and potential technology failures cannot be easily fixed.

Deciding on your testing location, then, should be a matter of what your resources allow you to do and what the goals of the test are. If you are creating an internationalized website (or product), then you want to test users from more than one location or country. Therefore, it may be more feasible to perform testing remotely. However, if you are creating a product specifically for a rural area in Kyrgyzstan, on-site testing may be a more reasonable choice. Your choices, as we have emphasized throughout the book, needs to be user and contextually centered.

10.8.2 TRANSLATION AND LOCALIZATION OF SCENARIOS AND TASKS

If you are testing users who speak a different language than you, you will need to work with a translator to make sure that you understand user feedback and your users understand scenarios and tasks. But more than this, scenarios and tasks need to be localized for your audience. Barnum (2011) notes that expecting products to be used in the same contexts in different cultures can confuse users when performing tasks. Barnum uses the example of a Chinese person withdrawing small sums of cash from an ATM, which is not a Chinese practice; therefore, asking a Chinese user to perform this type of task may result in confusion.

10.8.3 METHOD SELECTION

Research about international usability testing reveals a few patterns for you to consider when choosing methods for your test design. One such method is the think-aloud protocol (TAP). Barnum (2011), for instance, claims that TAP may be difficult for users when they are asked to speak in their non-native languages. However, Quesenbery and Szuc (2012) found that international usability testers preferred TAP, or "self-narrated tasks," to gather data (129).

Moderating the test and conducting structured interviews also show potential problems. In a study comparing how users interacted with a moderator from their own culture (Indian) versus another (Anglo-American), Vatrapu and Pérez-Quiñones (2006) found that Indian users interviewed by an Indian moderator were more open with their feedback and provided more culture-specific feedback than those interviewed by the Anglo-American moderator. Barnum (2011) also encourages moderators to be sensitive to ethnicity, age, gender, and status perceptions by other cultures (347), which may affect the way users interact with the moderator during testing.

Finally, self-reporting questionnaires, such as the system usability scale (SUS), have shown problems when used across cultures. Barnum (2011) explains that language may be confusing or difficult to translate when users are given these standard metrics. Further, Vatrapu and Suthers (2010), who used the questionnaire for user interaction satisfaction (QUIS) to gather data on user satisfaction, found that participants from China (who they describe as social collectivists) were more likely to rate satisfaction with the whole system higher than its parts. This was not the case of American (more individualistic) users.

This chapter demonstrates that international UCD ought to follow the same design pattern as any UCD project. You'll make additional plans, contexts, and accommodations to your design for international needs, but these additions are a matter of adjusting approaches and expectations to your research and design process.

TAKEAWAYS

1. Perform user field research to understand international users. If you cannot visit the users' locations, find partners who can. Remote research is also an alternative to traveling to your users' locales.
2. Investigate and design around users' cultural needs and their locations' infrastructures. Pay attention to economic and technological infrastructure, education and technology literacies, and devices available for technology use.
3. Choose internationalized or localized design based on the scope of your project.
4. Build and test design prototypes to learn about product interest and design preferences before engaging in costly international UCD.
5. An internationalization approach can create one product to serve the needs of many, varied users while saving organizational resources. The key to designing for internationalization is using the most generally accepted principle of design.
6. When choosing whether to internationalize or localize your product, the decision should come down to who the target users are and how users' goals can be met in the design.
7. If you have the resources and it makes sense for users, localization is a safer bet to satisfy user needs.
8. Users are different and deserve to be represented appropriately. Falling back on stereotypes to design for users devalues the UCD process.
9. If testing users who speak different languages, you should work with a translator to make sure that both you understand user feedback and users understand your scenarios and tasks.
10. Localize user scenarios and tasks for your audience.

DISCUSSION QUESTIONS

1. What are the barriers to locating international representative users? How might you address these?
2. How should a company decide whether to internationalize or localize a product? Isn't it too costly to design a product for so many different types of users?
3. What are some examples of internationalized or localized products? Do you use any internationalized or localized products? If so, how would you describe your use scenario?
4. Do acculturation and assimilation mean that designers will one day just be able to design one product that fits every user's needs?

REFERENCES

Barber, W., and A. Badre. 1998. "Culturability: The Merging of Culture and Usability." Proceedings of the 4th Conference on Human Factors and the Web, Basking Ridge, New Jersey, June 5, 1998: 1–10.

Barnum, C. *Usability Testing Essentials: Ready, Set...Test!* Boston: Morgan Kaufmann, 2011.

Bevan, N. 2009. International Standards for Usability Should Be More Widely Used. *Journal of Usability Studies* 4, no. 3 (2009): 106–113.

———. "International Standards for HCI and Usability." *International Journal of Human-Computer Studies* 55, no. 4. (2001): 533–552.

Callahan, E. "Cultural Similarities and Differences in the Design of University Websites." *Journal of Computer-Mediated Communication* 11, no. 1 (2005): 239–273.

Hall, E. T. *Beyond Culture*. Garden City: Anchor Press, 1976.

Hofstede, G. *Cultures and Organizations: Software of the Mind*. New York: McGraw-Hill, 2010.

Kulkarni, R., P. Rajeshwarkar, and S. Dixit. "Cultural Analysis of Indian Websites." *International Journal of Computer Application* 38, no. 8 (2012): 15–21.

Nielsen, J. "International Usability: Big Stuff the Same, Details Differ." *Nielsen Norman Group*. June 6, 2011, accessed February 26, 2016. https://www.nngroup.com/articles/international-usability-details-differ/.

———. "International Web Usability." *Nielsen Norman Group*. August 1, 1996, accessed February 26, 2016. https://www.nngroup.com/articles/international-web-usability/.

Quesenbery, W., and D. Szuc. *Global UX: Design and Research in a Connected World*. Waltham: Morgan Kaufmann, 2012.

Putnam, C., E. Rose, E. Johnson, and B. Kolko. "Adapting User-Centered Design Methods to Design for Diverse Populations." *Information Technologies and International Development* 5, no. 4 (2009): 51–73.

Rourke, C. 2013. "User-Centered Design Research for International Users." *Creative Bloq*. August 26, 2013, accessed February 26, 2016. http://www.creativebloq.com/netmag/user-centred-design-research-international-users-8135472.

Sharma, V. "A Comparison of Consumer Perception of Trust-Triggering Appearance Features on Indian Group Buying Websites." *Indian Journal of Economics & Business* 14, no. 2 (2015): 163–177.

Singh, N., and D. Baack. "Web Site Adaptation: A Cross-Cultural Comparison of U.S. and Mexican Web Sites." *Journal of Computer-Mediated Communication* 9, no. 4 (2004).

St. Amant, K. "Designing Web Sites for International Audiences." *Intercom* 50, no 5. (2003): 15–18.

Vatrapu, R., and M. Pérez-Quiñones. "Culture and Usability Evaluation: The Effects of Culture in Structured Interviews." *Journal of Usability Studies* 1, no. 4 (2006): 156–170.

Vatrapu, R., and D. Suthers. "Intra- and Inter-Cultural Usability in Computer-Supported Collaboration." *Journal of Usability Studies* 5, no. 4 (2010): 172–197.

Würtz, E. "A Cross-Cultural Analysis of Websites from High-Context and Low-Context Cultures." *Journal of Computer-Mediated Communication* 11, no. 1 (2005): 274–299.

Zhu, J. "4 Mistakes Behind Groupon's Failure in China." *Tech in Asia*. November 4, 2011, accessed February 26, 2016. https://www.techinasia.com/4-mistakes-behind-groupons-failure-in-china/.

11 Hardware UCD

Using the RABBIT process to create hardware is possible but presents different, often more complicated, multilayered challenges than software, web, or mobile applications of UCD. As Mike88 (2014) notes, "Hardware is hard for a bunch of reasons but one of the key reasons is its [sic] not just hardware that has to work" (sec. 2, par. 1). For a web development project, you may need to create HTML code, augmented by a particular scripting language, such as JavaScript, to promote dynamic interactivity. Or perhaps you need to write structure query language (SQL) queries to send or retrieve data from a database server. But for a hardware project, especially digital hardware, you likely need to do electronics customization, plastics modeling, device driver or software coding as well as software or web/mobile application coding to provide an interface for users to interact with the hardware.

After you create a prototype, it isn't enough just to test users interacting with a paper prototype of an interface. Hardware, once done, may be held or worn. Even if it is stationary, it possesses depth to its experience, requiring users to touch or manipulate it, and all of that demands testing. When you develop hardware, there is simply just a lot more going on, and that means more potential for problems, more need for others to participate, more money to spend to do it, more time to get it done—just more work and more challenges. In this brief chapter, we're going to spend a little time talking about a few best practices to consider when implementing RABBIT to develop hardware. As always, we'll rely on actual examples to illustrate our advice.

At EyeGuide, a decision was made in June of 2012 to transition away from its market-entry hardware platform to a more robust replacement that would be capable of carrying out more tasks, such as research or, in the case of the Focus product, concussion detection. The 1.0 hardware (Figure 11.1) was affordable but limited: Batteries for power were held in a small, square pack (worn across the front of the head) from which a movable arm, holding at its end an analog camera with an adjustable LED light, extended so that it could be placed over either eye to collect eye-tracking data. If users were patient, and if they couldn't afford a better system, they could make this hardware setup work.

But the camera literally had to be placed in front of the eye, which blocked users' normal vision. More importantly, there was no head movement compensation, so people wearing the hardware needed to keep their heads still during testing or the data collected would be poor if not unusable. The time it took to set up a user and the need to have some dexterity to manipulate the arm to get the camera into the right position, also meant, as Chapter 1 noted, that even though in theory this hardware could be used as part of a product for enabling users with limited or no hand functionality to control a computer mouse with their eyes, for all practical purposes, that was difficult if not impossible in reality. Given the need for some hand movement, as well as extended time to set up, the hardware was a weak link that made all applications but the most basic eye-tracking research unrealistic.

FIGURE 11.1   EyeGuide 1.0 Tracker. (Courtesy of Dr. Brian Still.)

EyeGuide developers, recognizing these limitations, made plans to innovate new hardware. Ongoing research and development led them to some important technological breakthroughs that gave them the confidence they could do it. They also analyzed the marketplace and received substantial feedback from users (namely customers) during the first 18 months of the company's existence and its deployment of the early generation hardware. All of this information, the research (Chapter 4) on users and the assessment of the situation (Chapter 5), was used to create a balanced set of requirements (Chapter 6) for what was to become EyeGuide's next generation hardware (Figure 11.2).

To maximize competitive white space in multiple markets, ranging from video game play to drone and robotics control, to sports, driving, and other in-context mobile research as well as, of course, medical treatment (i.e., Focus), the new EyeGuide 2.0 hardware, it was decided, needed to have the following:

- Comfortable wearability and flexibility for multiple, different-use situations (driving in a car, playing sports, playing video games, carrying out medical testing, etc.)
- Long use lifetime in the field via battery power
- Backup storage

FIGURE 11.2   EyeGuide 2.0 EG hardware. (Courtesy of Dr. Brian Still.)

- Wireless connectivity for monitoring what EG wearers were seeing and doing
- Platform independence so that EG could be made to work with a variety of different products, operating systems, and other devices
- Control capability through an eye camera as well as an accelerometer that captured head movement
- High-quality scene camera to capture HD, wide-angle, realistic views of the world through the wearer's perspective

To make all of this work in a single system and to be user-friendly, EyeGuide followed the RABBIT process, creating a series of prototypes for users to test.

Building an operative image (Chapter 7) through different stages of fidelity, iterating newly designed prototypes after feedback at each stage from users, held high value for EyeGuide developers, and it was something they followed. But the context they worked in to carry out the development and testing of prototypes, because it was hardware, was, as already noted, altogether more challenging.

11.1   TEAM DYNAMICS IN HARDWARE DEVELOPMENT

For one, it is rarely the case that a single person or even a single team of similarly skilled developers can create hardware, especially digital hardware, such as the EyeGuide eye-tracking hardware. Think about it for just a second. What powers the hardware? Batteries, right, but the batteries are just the gasoline. Most cars need gasoline to drive, but the gasoline is put to use by an engine, and that engine is told what to do, these days at least, by a computer or set of computers that manages everything the car does. This is also true for digital technologies, such as EyeGuide's eye tracker. Printed circuit boards (PCBs) are tiny electronic engines that make everything in the hardware work and put the energy of the batteries to effective use. An electrical engineer trained in PCB creation needs to be on a hardware team. So does a mechanical engineer or industrial designer who has knowledge of how to create plastics or other encasing materials (wood, metal, etc.) to house PCBs to fit well in the environment where they will be used or worn and also carry out the required functions that users need and want.

Other electrical engineers and software developers have to create software, which users never see, to tell cameras, speakers, and other devices what to do. Then there are sometimes ergonomic experts, graphic designers, and, of course, software or web/mobile application developers. The point is a hardware development team by necessity is pretty large. You don't need to have all of these people on your team, but that means you will face the challenge of managing vendors or outside contractors who work for you. They are sometimes working on other projects that distract them, or they don't care as much as you do about what you're trying to accomplish. They can also come from different cultures, organizational or even country-based. At various points, EyeGuide had mechanical engineers in Minnesota designing plastics, electrical engineers in India designing PCBs, metal industrialists in California making headbands, leather manufacturers in China making ear caps, even optics engineers in Oklahoma making hot mirrors to capture a reflection of the eye camera's image of the eye so that users' vision would not be obstructed by the camera.

Your hardware project may be simpler, and you may choose to do cruder fabrication, making use of off-the-shelf parts or other workarounds in your earliest stages to avoid the complications and accompanying expenses of a large team. In fact, we encourage you to take a "junkyard" approach early on. Use a wood block when you don't have access to high-quality plastics. Repurpose an existing set of headphones, as EyeGuide did at the beginning of its efforts, to stand in as a prototype of its proposed hardware (Figure 11.3).

The EyeGuide team even put web cameras in place of where eye and scene cameras would at some point go because developers wanted to know fit and location by testing with real users before they spent the time and money to engineer customized solutions.

But, at some point, if you develop hardware that you want to bring to market, you're going to need to manage a diverse team of people, including vendors from other organizations and cultures. Make them work together. As Mike88 (2014) writes, "One of the mistakes I made was not forcing the hardware and firmware engineers, the mobile app developer and the industrial designer to sit alongside each other and look each other in the eye all day every day" (sec. 2, par. 3). EyeGuide set aside a complete room, without cubicles, and forced its developers to work with each other on the same tables, using the same tools in the same space. Virtually, it set up Skype sessions and had vendors working offsite, such as in India, to call in and participate. On more than one occasion, EyeGuide even paid for the expenses of an electrical engineer to come on location to EyeGuide headquarters and work during a crucial stage of prototype development.

FIGURE 11.3 Early version of EyeGuide hardware that makes use of existing headset, in-house wiring, and crude plastics molding to work. (Courtesy of Dr. Brian Still.)

Finally, require stand-ups every day just like agile software. There are no stupid questions or bad ideas, just opportunities for design to be tested with users. Also, don't ever use cost as an excuse during prototyping. You can test without much expense, and spending the time and money to reiterate a prototype early on saves exponentially on changes later.

11.2 MATERIALS, TOOLS, AND COSTS

Plastic parts, leather pieces, cameras, speakers, accelerometers, LED lights, screws… when you create hardware, there will be a number of components. Even seemingly simple hardware devices have more components than you realize. On top of that, you will need to have tools on hand or find a vendor that has them to make even prototypes of hardware. In the last few years, there has been a revolution in agile, fast hardware prototyping because of the creation of low-cost rapid 3-D plastics printing machines. EyeGuide didn't have one for its 1.0 hardware, but for its 2.0, it bought a Makerbot Rapid Printer, enabling developers to experiment quickly with different prototypes of the 2.0 eye-tracker hardware. The team created a model using a 3-D modeling software called AutoCAD, and then that model was loaded into the rapid printer. Depending on the complexity of the model, parts could be produced in a day. EyeGuide would frequently produce a new iteration of the headset to check for fit on a user's head or to determine the position of cameras.

One of the biggest challenges, in fact, for developers, led ultimately to a patent. The camera arm had to be positioned so that the hot mirror captured all of the user's eye in the reflection from the eye camera, but at the same time, the scene camera, also positioned on the camera arm, captured, in turn, an adequate, realistic view of what the user was looking at in the real world while wearing the eye tracker. The challenge was immense, but thanks to the rapid 3-D plastics printer and a lot of user testing of prototypes, EyeGuide finally found an optimal solution.

Hardware development, by nature, requires more materials, more tools, and thus more money. You could literally build a web application for next to nothing, assuming you had affordable or free access to a web server, scripting software, and some device to program on as well as upload files and monitor activity. Not so with hardware. Cheap but good rapid printers are hundreds if not thousands of dollars, not including parts and materials. PCB fabrication is expensive. There are inexpensive starter kits, and you can do your own soldering, but, at some point, you can't sell an off-the-shelf hardware device that is a Frankenstein version of other people's work unless you plan to pay them for the rights, which costs money. And if you don't have the skills yourself, you'll have to pay other people. There's no getting around it. By the time EyeGuide finished making just 250 of its 2.0 eye-tracking product, it had spent almost $500,000 for everything, from research and development, prototype creation and testing, parts procurement, labor, production, assembly, and finally testing.

11.3 PROTOTYPE ITERATION AND TESTING

Mike88 (2014) writes, "Be prepared to throw your prototype away and start again" (sec. 6, par. 1). You can't use the excuse that because you are creating hardware that you cannot prototype as much. In fact, you need to prototype hardware just as much, if not more, than software because the challenges for your product extend beyond a digital interface. Knobs, handles, buttons, and other form factors, such as weight, size, and wearability, all have to be prototyped in addition to the same things you're interested in evaluating for software.

Think of Garrett's (2011) layers again (Chapter 2). We know we need to account for a lot of important things that occur below the surface of a product's interface, and when it comes to hardware, the depth and complexity of scope, strategy, skeleton, and structure layers underneath the surface are challenging. Remember the alarm clock example from Chapter 3. If you don't give enough consideration to how or where the alarm clock will be used and who will use it, such as business travelers in a hotel who want fast and easy, then at the strategy and subsequent skeleton and structure layers you'll allow for far too many buttons, too many options on the clock that overwhelm and frustrate users.

Don't make assumptions about the hardware that cause you to skip basic form and feel to get to more advanced, hi-fi prototypes. Resist the urge telling you that you need to have more finished hardware before you begin testing. Prototype a lot. Prototype early. Prototype quickly. You can start with a plain white box if you're building a clock or maybe a wooden block for a smart device. Douglas (2012) points out that even pen and paper sketching has value: "Simple sketches can help realize ideas and form them to guide physical prototypes. There is often a lot of different ways to build or do something. Having different ways on paper can help deciding which direction to take. They can also express your ideas quickly to other people" (sec. 3, par. 2).

In EyeGuide's case, developers knew they wanted a headset but weren't sold on a particular design at the beginning. Ear-caps? Glasses? Because developers wanted user involvement in the process from the beginning, developers experimented with a variety of prototypes. The team had access to a rapid printer, so it could create prototypes with higher fidelity, but that can be worked around if you lack such resources. For the 1.0 eye tracker, the EyeGuide team didn't have a rapid printer, so it bought off-the-shelf shop glasses and also modified an existing set of headphones. Users were tested using both, so the team could decide which design was best to pursue going forward.

Ultimately, PCBs, cameras, and other electronics have to be designed to fit in the plastics or materials that make up the product. If you rush through your decision on the form factor of the materials, you may create electronics to fit a design that doesn't work. If you think it's expensive to change a software interface late in the process, after it is more advanced in its development, just imagine the costs of changing the size of the electronics to fit a new hardware design—they are enough to put you out of business.

11.4 LOW-FIDELITY, PARTIAL PROTOTYPES ARE JUST FINE

Paper, wood, anything you have access to and that you have the ability to shape into a recognizable form factor resembling your hardware's intended look and feel can be used for prototyping. At one point, EyeGuide developers bought extra thick construction paper, cut out a shape from it that looked like a headset, and then put it on different users to get a sense of where the eye and scene cameras would go. You can see that they also bought sunglasses, cut out the glasses from the frames, and then mounted a brass bar with a web camera to one side of the frames to test eye camera location (Figure 11.4). Even without money, you can be creative in prototyping hardware. Find workarounds. In fact, just forcing yourself to come up with early-stage prototypes forces user feedback early as well as a proactive conversation among team members that often identifies and solves problems in design that can be addressed sooner rather than later.

Sylvia Wu (n.d.) notes that a rougher prototype that lacks "high fidelity color, material, and finish can actually be beneficial: it forces...[the user] to concentrate on the form of the product. Form helps to convey how something might be used, and you need to confirm that your design provides affordance effectively to the user" (sec. 2, par. 11). Less is more. In fact, don't feel the need to create the entire form. Wu argues you can just create modules or parts of the form (such as the seat of the bicycle and not the entire bicycle) and still learn quite a lot about what users think about the larger product. EyeGuide developers learned, for example, with its rough, construction paper prototype testing, that a headset design would work best for client and user needs because it better accommodated glasses wearers. An eye tracker in glasses form meant people who wore glasses couldn't wear the eye tracker, and that eliminated a significant population. Anyone could wear a headset.

The images in Figures 11.4 through 11.8 show hardware prototypes of varying fidelity, which EyeGuide developers tested with user feedback, attempting, in the

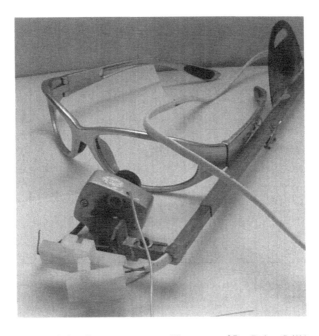

FIGURE 11.4 EyeGuide hardware prototype. (Courtesy of Dr. Brian Still.)

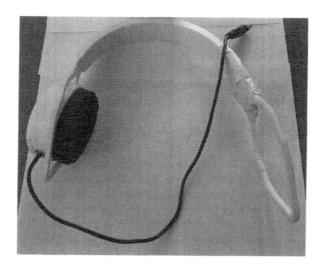

FIGURE 11.5   EyeGuide hardware prototype. (Courtesy of Dr. Brian Still.)

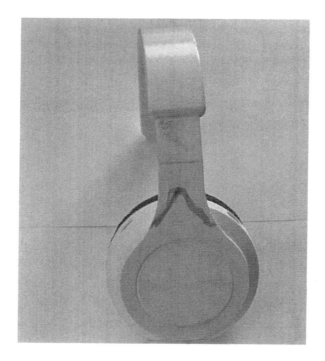

FIGURE 11.6   EyeGuide hardware prototype. (Courtesy of Dr. Brian Still.)

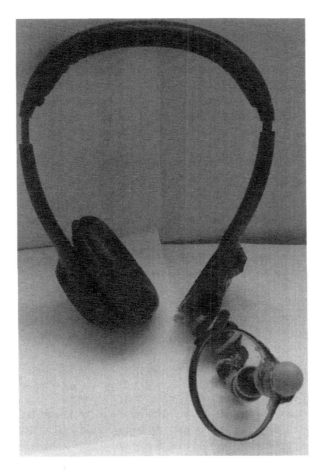

FIGURE 11.7 EyeGuide hardware prototype. (Courtesy of Dr. Brian Still.)

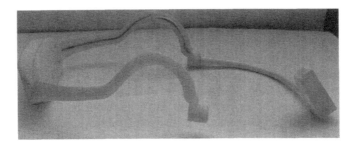

FIGURE 11.8 EyeGuide hardware prototype. (Courtesy of Dr. Brian Still.)

process, to find a design that both worked for the product's intended users and delivered on the company's aforementioned requirements. Of these, Figure 11.6 shows most obviously the iterative nature of RABBIT hardware development. You can see clearly where EyeGuide developers made markings to indicate changes required to improve the headset fit after user testing. The prototype headset wasn't flexible enough to fit a wider range of head sizes. Developers learned this quickly and sketched in pencil the adjustments necessary to correct this in a subsequent prototype that was then printed rapidly and again tested. Overall, the rapid production and subsequent testing of this prototype shown in Figure 11.6 and, for that matter, the creation and testing of other prototypes allowed EyeGuide to learn quickly, build quickly, test quickly, and improve quickly.

11.4.1 AVOID TOO MUCH WHITE SPACE

With hardware especially, having too much competitive white space can be damning. You don't want to make another Apple Lisa or Ford Edsel. You want to be innovative but not at the expense of your targeted users being able to use your hardware effectively. Look at what competitor products or industry leaders do and find ways to create form and feel that takes advantage of their successful traits or at least mirror size, weight, and other factors that users have become accustomed to using. Too much white space may mean users can't use your hardware or at least can't use it effectively.

11.4.2 GET OUT OF THE BUILDING (GOOB)

Lanoue (2015) argues you should test hardware prototypes in a lab or other controlled environment: "When you're testing a physical prototype and only a few of them exist, it's probably best to bring participants into the lab. That way you can keep the prototypes in your possession and moderate your test participants" (sec. 1, par. 3). There is, of course, some logic to this, but we believe you need to get out into the field and test to gather the best feedback from your users. Find a way to create multiple versions of the prototype, if possible, get users to sign a nondisclosure agreement, and, if privacy is still a concern, figure out other ways to keep your real-world testing as unnoticeable as possible. You need to test in the field. Remote testing makes moderation of users difficult, but it can be done. Users will try to make your hardware work in the lab, or they will change their behavior because they are testing the product in the lab. But in the real world, the hardware's fit with users' environments will be exposed immediately through more realistic user behavior. It will be a challenge, but find ways to go where users work, live, and play to see what kinds of hardware, similar to your proposed product, they are using and how they are using it. Then, take your prototypes out into the field and test there.

11.4.3 REMEMBER THAT SIMPLE IS BETTER

Recall Colborne's (2011) notions of simple and usable design mentioned in Chapter 3. Resist adding too much—too many buttons, too many knobs, too many instruments; too many and too much of everything overwhelms users both mentally and physically. Think clean, displaced, simple design. Try to accomplish more with less, and only add features as warranted through user feedback during testing.

11.4.4 Make Sure You Have Enough Stuff

Early on, order enough components, such as cameras and other parts that go into your hardware prototypes, so that you can continue to test and build after some are broken or lost. If you think you need three cameras, buy six. If you don't know what camera to use, then buy a few different kinds.

It is easier to buy components from U.S. suppliers if you only need a few. They ship relatively fast in most cases. However, they are more expensive because they buy from overseas suppliers and then mark them up to make a profit. You can skip these suppliers and go directly to manufacturers overseas, but (1) they often take a long time to send components and (2) they make their money on bulk orders. If you want them to give you components for testing, you'll have to negotiate with them to supply samples and promise often to buy in bulk later on.

11.4.5 Use Tools of the Trade for Higher Fidelity Prototypes

You don't need a 3-D rapid printer for plastics, but if you're serious about this, it wouldn't hurt to have one at some point. You can pay for vendors to make molds and plastics for you in small runs, but you might find that more expensive than doing it yourself using your own printer. You also can skip this by going to a local hobby store (on location or online) and purchasing chemicals for mixing plastics that you pour and set into your own handmade molds: cheaper but cruder. EyeGuide used this often for its 1.0 product.

Finally, if you're going to use electronics, you'll likely need to purchase and become relatively skilled at using a soldering kit (or bring on to your team someone who can use it). When you are prototyping electronics, you must invariably create wired connections to power your custom or off-the-shelf components.

REFERENCES

Colborne, G. *Simple and Usable: Web, Mobile and Interaction Design.* Berkeley: New Riders Publishing, 2011.

Douglas, G. "How to Build It: Lean Prototyping Techniques for Hardware." *Graeham Douglas.* November 15, 2012, accessed February 25, 2016. http://graehamdouglas.com/2012/11/15/lean-prototyping-for-hardware/.

Garrett, J. *The Elements of User Experience: User-Centered Design for the Web and Beyond.* Berkeley: New Riders Publishing, 2011.

Lanoue, S. "The 3 Most Common Ways to User Test Your Product." *User Testing Blog.* May 6, 2015, accessed February 25, 2016. https://www.usertesting.com/blog/2015/05/06/3-approaches-for-collecting-user-feedback/.

Mike88. "Prototyping Hardware – 15 Lessons Learnt the Hard Way." *Startup88.* September 30, 2014, accessed February 25, 2016. http://startup88.com/hardware/2014/09/30/prototyping_hardware_15_lessons_learnt_the_hard_way/2336.

Wu, Sylvia. "Strategic Testing Methods for Early Stage Hardware Development." Fictiv. n.d., accessed February 25, 2016. https://www.fictiv.com/resources/starter/strategic-testing-methods-for-early-stage-hardware-development.

12 Print UCD

The term "print" refers to a centuries-long tradition of paper-based communication (such as correspondence, record keeping, and information delivery). Until the late 20th century, paper communication dominated most organizations' record-keeping and archival systems. Although today, in a more digital world, we have technology that can streamline such processes and allow for our communication to be more digital, we still use print communication to distribute surveys and questionnaires; instructions and instruction manuals for products; and registration, patient history, and feedback forms. We even still read print newspapers, magazines, and other periodicals. Print still matters, whether that means off-line, hard-copy print documents we hold and read or print documents made available for electronic distribution, often called PDFs, which stands for portable document format.

Although our focus in UCD is often on websites or software, UCD for print documents must be discussed. In this chapter, we take advantage of the RABBIT framework to provide an overview of how to design printed documents for users, covering how to research the users' environments to determine when print or PDF materials are needed to complete tasks, balance the needs of users with the requirements of your client or organization, create MVPs and prototypes specific to the purpose of the project and the print-mode of delivery, and test for simple and complex tasks using print materials.

12.1   RESEARCHING USERS

You already know that observing users' current uses of similar products in their environments is necessary to understand how new products can or ought to fit into their current systems of use. For example, you could have the best, most beautiful web design in mind for providing community information on water quality. However, if users of this information lack the resources for viewing the web page or do not have the web literacy necessary for finding information on their community's water quality, you have done the community a disservice despite your best intentions. The same principles apply to how information will be used. Redish (2010) argues, "Whatever you are creating (hardware, software, web application, information website, e-commerce website, stand-alone document), the product would have no reason to exist without users. The interaction and interface of your product are the ways that the product 'affords' itself to its users, or how it communicates with its users" (197).

The fire blanket in Figure 12.1 is an example of how the communication "affords" itself to users. It has clear instructions on the outside of the packaging. The instructions explain how to use the blanket for two situations: extinguishing kitchen fires or clothing fires. The instructions use clear images, text, and color to make reading and using information easy. Imagine if you had to Google instructions on how to

FIGURE 12.1   Fire blanket. ("Fire blanket," own work, Praewnaaaaaam, June 2, 2015. Creative Commons attribution 4.0, https://creativecommons.org/licenses/by-sa/4.0/deed.en.)

use the fire blanket; you would put yourself and those around you in further danger. You need to make decisions about the mode of delivery based on the context of your users.

When designing information, whether print or digital, understanding the users' use of a document ought to inform how the document is designed and delivered. Porter (2008) argues that the art of creating information is "degraded when it is taught or practiced as a set of mechanical procedures, rules or formulas to be followed or patterns to be copied. It achieves status as a true art when it [is] taught and practiced as [a] form of knowledge involving a critical understanding of the purposes and effects of the art on audiences and the practical know-how to achieve those effects in new discursive situations" (5). In UCD terms, your information design and delivery decisions should derive from the audiences' (or users') contexts, and decisions are reassessed as users and their environments change. For both the water quality information and the fire blanket instructions, user research, such as surveys about user demographics, access to and use of technology, and preferences for document delivery; interviews with users about their use of documentation; and existing documents that will be redesigned, would yield helpful results. Further, understanding the environment (or where the document will be used) is essential to choosing the right mode of delivery, information needs, and design.

12.2   ASSESSING THE SITUATION

Once you know how users will use your document, it is time to consider your design options. This is also a time to collect the list of requirements needed by both your client or organization and your users. Let's look at an example of how one designer used user and organization information to determine the best design for a seminar registration form.

A designer took a new position as assistant coordinator of an annual two-week seminar for an academic program. One of her first tasks was to create a new registration form for the event. People registering for this event were located all over the world. Thus, the form needed to be sent out electronically. The form also needed to be sent in a format that would allow users to complete it without needing a special software program. However, because the seminar was meant for students at the university she worked for, she was unable to publish and collect an online form due to the Family Educational Rights and Privacy Act (FERPA). Under FERPA, you cannot collect students' personal information using unsecured or third-party websites. Therefore, the form had to be delivered and submitted through secure channels: either postal mail or the organization's email server.

The previous years' registration form was designed, sent, and returned as a Word document. However, this year, the seminar coordinator wanted to experiment with a PDF form that would allow users to fill in details that then would be aggregated into a single spreadsheet collecting all of the respondents' details. This meant that a PDF needed to be created with the expectation that all users would use Adobe Acrobat Pro (or another software workaround) to fill out and save the form. The coordinator also had a list of information the designer needed to include on the form, which included students' personal information, faculty advisors, seminar year, presentation or workshop details, and dietary restrictions.

Therefore, the goal for the new design was to create a secure, PDF form that could be completed using Adobe Acrobat Pro. Further, it had to include the coordinator's requested information fields to collect necessary data from participants.

12.3   BALANCING USER NEEDS

When working to balance user needs for printed materials, the process does not change from other types of design. The first step is to gather the data about users' environments, workflows, resources, and knowledge. Then, using that data, you create profiles for the document's users and use. Once these steps have been accomplished, you can begin working on a design that will balance the needs of both your various users and the organization.

In the seminar registration form example, users of the seminar registration form were widely dispersed geographically; thus, the normal means for communication with seminar participants and administration was email. A listserv for this group of participants had been long established. Because this was a direct line of communication, the coordinators planned to create a PDF form and distribute the form via the email listserv, providing information on how to fill out and return the form within the email content (the registration form was attached as a PDF file). Then, participants were asked to fill out the form, attach it to an email, and send it back to the assistant coordinator.

Email communication and electronic PDF forms were also the best means for the event coordinators. The PDF could be created fairly easily using existing forms from previous seminar years and transformed into a PDF form using Adobe Acrobat Pro, which the assistant coordinator had access to. Receiving the completed forms also meant that the assistant coordinator could automatically generate data into one data spreadsheet as the coordinator requested. However, participants had to fill out the form and save it in Adobe Acrobat Pro, not Adobe Reader. The only way users could save the form using Adobe Reader was to "print" the file to "save as PDF." When data was exported to the spreadsheet, the "save to PDF" form acted like a photocopy and form field information could not be extracted to the spreadsheet. The need to save the form in Adobe Acrobat Pro was written in the instructions of the email message.

To balance user needs, the designer needed to use users' existing means of communication for distributing the form and also rely on users having access to and knowledge of Adobe Acrobat Pro to complete the form.

12.4 BUILDING AN OPERATIVE IMAGE

It may seem odd to create prototypes of print documents. Often, when it is time to create print documents (whether textually based or graphic design prints), you begin by sitting in front of your computer and creating. However, it helps to first map out content and get feedback. If you can create prototypes with users, you can get even more insight into users' mental models.

When designing websites, many designers understand that a strategy for information architecture and interface design is as necessary as content. Print is not an exception to this rule. Finding ways to create usable organization for navigating content, for instance, is imperative for any document. The study of information and document design exists for this very reason. Headers, spacing, color, and typography are all used to create designs that make content usable. In fact, many principles of document design (such as contrast, repetition, alignment, and proximity) expect that similar uses of design principles connect with users' mind models for better document use. In Durack's (1997) study of sewing patterns, she notes that contemporary sewing patterns "are substantially identical in key ways and represent a single genre of technical communication" (50). Genre and design principles can help in understanding the expectation of users, but it cannot replace actual user feedback.

User input at this stage of design can (and should) take the guesswork out of design. Yes, we have great design principles to follow (such as principle of contrast, repetition, alignment, and proximity), but how these principles connect to user needs for completing a task, such as filling out a form or finding information to complete a task, may not be as clear as the designer thinks.

One way to begin thinking about design is to create an MVP, or minimum viable product, to see if your plan for document delivery or content is helpful for users. If you remember from Chapter 7, the MVP can take different forms. Its goal is to get feedback from users as early as possible to help guide your efforts in creating a more user-centered prototype.

For the seminar registration form, the designer's MVP was based on two simple questions: (1) Do you have experience filling out Adobe forms, and (2) do you have access to Adobe Acrobat Pro? The answers to these two questions helped confirm assumptions the designer and event coordinator had about using a PDF fillable form and helped them, in the process, create instructions on how to fill out and save the form properly. With data gathered from the MVP, the designer compiled a checklist of content that would be required in a paper prototype (Figure 12.2).

Major content	Information needed
Personal information	First and last name
	Student identification number
	Cell phone number
	Permission to disclose cell phone number to faculty and participants
	Name of emergency contact
	Phone number for emergency contact
Dissertation/annual review committee	Indicate if you have a dissertation committee
	Provide names of chair (if available), committee members, or faculty you would like in your annual review (if you don't have a committee).
Seminar presentation	Use presentation descriptions from the website to create selection of seminar presentation types based on year of attendance.
	Presentation title
	Presentation abstract
	Technology requests for presentation
	Print requests for presentation materials
Meals and events	Dietary restrictions
	Farewell dinner attendance
	Farewell dinner guest
	Vegetarian option for farewell dinner
Additional options	Rec center pass
	Parking permit
	Anything missing? Please include this information in your prototype.

FIGURE 12.2   Paper prototype checklist of needed content.

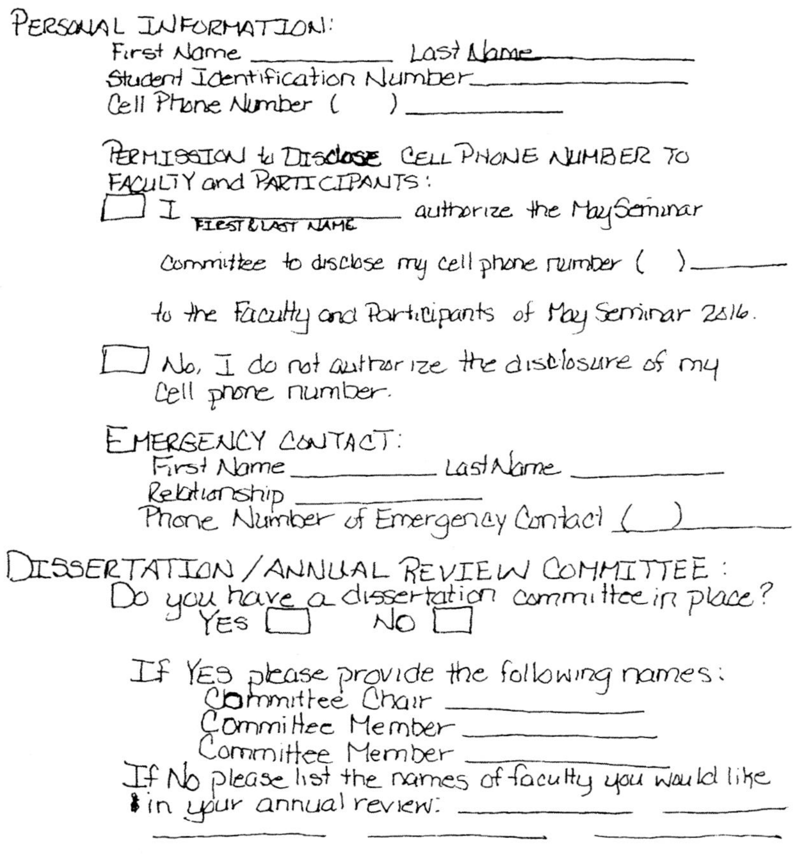

FIGURE 12.3   Form prototype page 1. (Courtesy of Dr. Kate Crane.)

The paper prototype derived from this checklist, seen in Figures 12.3 and 12.4, allowed the designer to put a low-fidelity version of the registration form in front of users (seminar attendees), who then provided their feedback on what they would expect to see in the form, such as information to be displayed and organized.

This early stage back-and-forth with users allowed the designer to understand the model the users would expect to see. From there, the designer created a more advanced, hi-fi prototype with fillable form fields that she then used for additional testing.

12.5   TESTING

Print documents can be tested for information to complete tasks as well as how information provides knowledge to complete tasks. Simple task-based testing can be thought of as having a task that has a clear start point and a clear end point

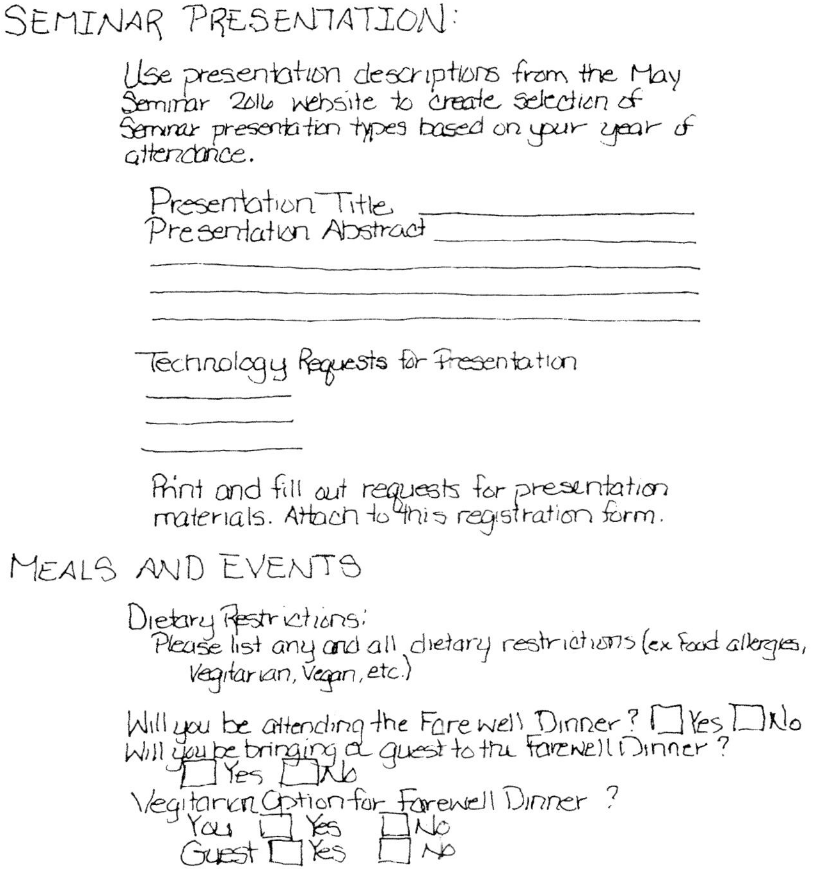

SEMINAR PRESENTATION:

Use presentation descriptions from the May
Seminar 2016 website to create selection of
Seminar presentation types based on your year of
attendance.

Presentation Title _____
Presentation Abstract _____
_____
_____
_____

Technology Requests for Presentation
_____

_____

Print and fill out requests for presentation
materials. Attach to this registration form.

MEALS AND EVENTS

Dietary Restrictions:
   Please list any and all dietary restrictions (ex food allergies,
   Vegitarian, Vegan, etc.)

Will you be attending the Farewell Dinner? ☐ Yes ☐ No
Will you be bringing a guest to the Farewell Dinner?
       ☐ Yes ☐ No
Vegitarian Option for Farewell Dinner ?
   You   ☐ Yes   ☐ No
   Guest ☐ Yes   ☐ No

FIGURE 12.4   Form prototype page 2. (Courtesy of Dr. Kate Crane.)

(as discussed in Chapter 8). Instruction documents (such as the fire blanket instructions) or registration forms are examples of these task-based tests.

Documents may also be more complex in nature and need to be tested not only for finding information to use for a task, but also for testing the content/information itself and how it is used to complete a task. Albers (2011) calls these tasks complex, which means there are no clear beginning or ending points to the task. The 1040EZ tax form (Figure 12.5) is an excellent example of a document used for a complex task.

There are different ways to test these two types of tasks. For both types of tasks, you should use the *see-say-do triangle*. The one difference for testing complex tasks is that you want a measurement that shows not only how users find information, but also how they use the information to complete tasks. Thus, in creating tasks and analyzing data, you want to incorporate a request for using the information. For instance, if testing the tax form, you might ask, "How was information about earned

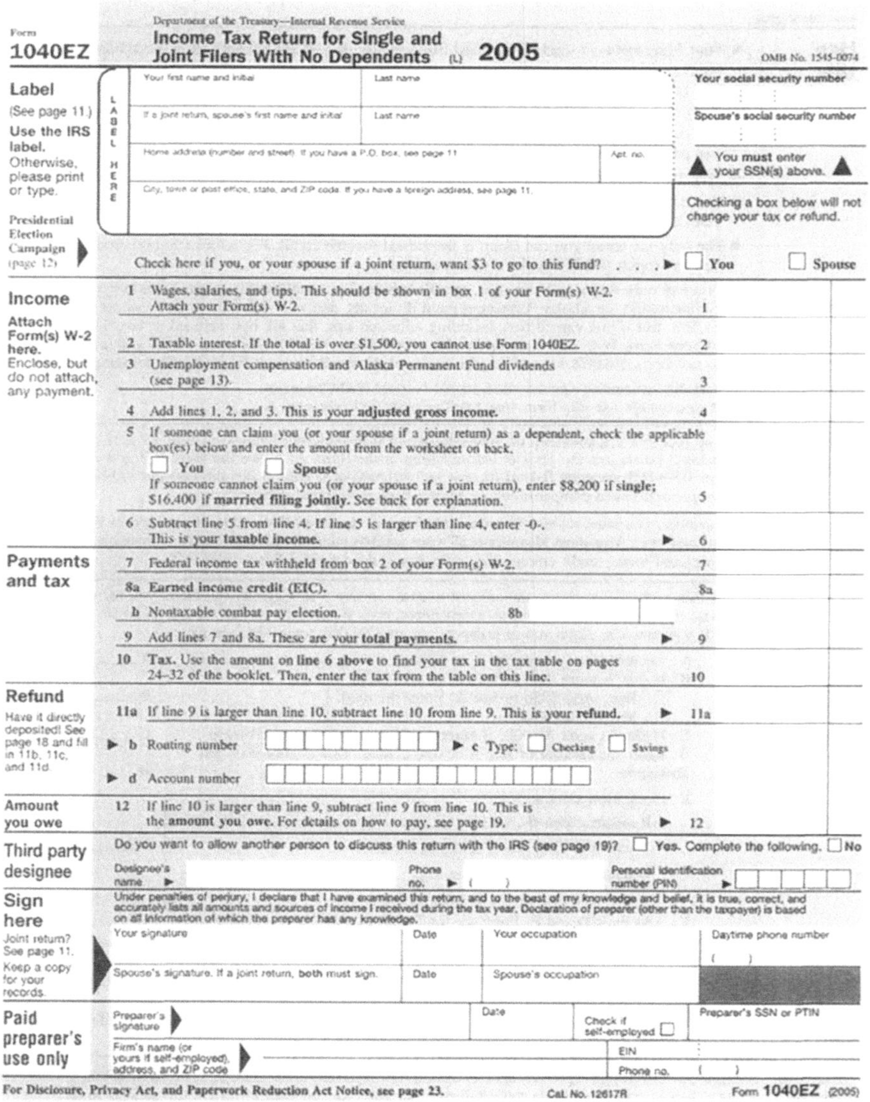

Form 1040EZ — Department of the Treasury—Internal Revenue Service
Income Tax Return for Single and Joint Filers With No Dependents (L) 2005 OMB No. 1545-0074

Label (See page 11.) Use the IRS label. Otherwise, please print or type.
LABEL HERE
Your first name and initial | Last name | Your social security number
If a joint return, spouse's first name and initial | Last name | Spouse's social security number
Home address (number and street). If you have a P.O. box, see page 11. | Apt. no. | ▲ You must enter your SSN(s) above. ▲
City, town or post office, state, and ZIP code. If you have a foreign address, see page 11.
Checking a box below will not change your tax or refund.

Presidential Election Campaign (page 12) ▶ Check here if you, or your spouse if a joint return, want $3 to go to this fund? . . . ▶ ☐ You ☐ Spouse

Income
Attach Form(s) W-2 here.
Enclose, but do not attach, any payment.

1 Wages, salaries, and tips. This should be shown in box 1 of your Form(s) W-2. Attach your Form(s) W-2. — 1
2 Taxable interest. If the total is over $1,500, you cannot use Form 1040EZ. — 2
3 Unemployment compensation and Alaska Permanent Fund dividends (see page 13). — 3
4 Add lines 1, 2, and 3. This is your **adjusted gross income.** — 4
5 If someone can claim you (or your spouse if a joint return) as a dependent, check the applicable box(es) below and enter the amount from the worksheet on back.
☐ You ☐ Spouse
If someone cannot claim you (or your spouse if a joint return), enter $8,200 if **single;** $16,400 if **married filing jointly.** See back for explanation. — 5
6 Subtract line 5 from line 4. If line 5 is larger than line 4, enter -0-. This is your **taxable income.** ▶ 6

Payments and tax

7 Federal income tax withheld from box 2 of your Form(s) W-2. — 7
8a **Earned income credit (EIC).** — 8a
b Nontaxable combat pay election. — 8b
9 Add lines 7 and 8a. These are your **total payments.** ▶ 9
10 **Tax.** Use the amount on **line 6 above** to find your tax in the tax table on pages 24–32 of the booklet. Then, enter the tax from the table on this line. — 10

Refund
Have it directly deposited! See page 18 and fill in 11b, 11c, and 11d.

11a If line 9 is larger than line 10, subtract line 10 from line 9. This is your **refund.** ▶ 11a
▶ b Routing number ▶ c Type: ☐ Checking ☐ Savings
▶ d Account number

Amount you owe

12 If line 10 is larger than line 9, subtract line 9 from line 10. This is the **amount you owe.** For details on how to pay, see page 19. ▶ 12

Third party designee
Do you want to allow another person to discuss this return with the IRS (see page 19)? ☐ Yes. Complete the following. ☐ No
Designee's name ▶ Phone no. ▶ ( ) Personal identification number (PIN) ▶

Sign here
Joint return? See page 11. Keep a copy for your records.
Under penalties of perjury, I declare that I have examined this return, and to the best of my knowledge and belief, it is true, correct, and accurately lists all amounts and sources of income I received during the tax year. Declaration of preparer (other than the taxpayer) is based on all information of which the preparer has any knowledge.
Your signature | Date | Your occupation | Daytime phone number ( )
Spouse's signature. If a joint return, both must sign. | Date | Spouse's occupation

Paid preparer's use only
Preparer's signature ▶ | Date | Check if self-employed ☐ | Preparer's SSN or PTIN
Firm's name (or yours if self-employed), address, and ZIP code ▶ | EIN | Phone no. ( )

For Disclosure, Privacy Act, and Paperwork Reduction Act Notice, see page 23. Cat. No. 12617R Form **1040EZ** (2005)

FIGURE 12.5 1040EZ fillable PDF tax form, public domain. The tax form is usually filled out after separate worksheets are completed to find information (or handy software programs that calculate your taxes more easily).

income credit under 8a used to determine users' eligibility for the earned income credit?" This means users would have to not only find information about the earned income credit, but also use that information to determine their eligibility and incorporate that decision into completing their 1040EZ form.

For the seminar registration form, five veteran seminar participants tested the PDF hi-fi prototype (Figures 12.6 through 12.8). They were asked to fill out the

May Seminar 2013 Registration Form

Please complete this registration form, save it as a pdf, and return it to Kate Crane (kate.crane@ttu.edu) no later than April 1, 2013.

Personal Information

First Name: _____   Last Name:_____

Student ID/R#:_____   Cell Phone#:_____

Do we have permission to release your cell number with May Seminar attendees?

◯ Yes
◯ No

Emergency Contact Name:_____

Emergency Contact Phone:_____

Dissertation/Annual Review Committee

Do you have a dissertation committee?

◉ Yes
◯ No

Please list your committee members below. If you are new to the program or have not formed your committee, you do not need to complete this section; however, if you have faculty in mind that you would like on your annual review, please fill in those names. We may not be able to accommodate all requests, but we will try.

1.   Chair:_____

2.   _____

3.   _____

Seminar Presentation Information

See presentation descriptions for more information about presentation formats. Please note, 1st year students only need to fill out the May Seminar Year in this section—no title, abstract, or special technology/printing information is needed for Rapid Rhetoric.

May Seminar Year:

◯ 1st Year—Rapid Rhetoric
◯ 2nd Year—Individual Conference/Panel Presentations (15-20 minutes/person)
◯ 3rd Year—Poster Presentation
◯ 4th Year—Graduate Research Network
◯ 5th Year—"Job Talk" (30-45 minute talk) or Dissertation Defense (with approval from chair/committee)
◯ I will not be presenting because I am taking qualifying exams or writing dissertation (upon chair/committee approval only)

◯ I will not be attending this year's May Seminar because I have attended my required 5 May Seminars or because I have been given exemption by my chair for personal reasons.

FIGURE 12.6   May seminar event PDF fillable form page 1. (Courtesy of Dr. Kate Crane.)

form using their own information and provide feedback on document errors or difficulty of use. This was not a formal test, rather one that assumed users would find information representative of the larger group of users through their own knowledge of the registration process. Users used their own data and equipment to complete their forms; in this case, completing the form in their own environments provided the best means of evaluation. The designer needed to learn about

Title of Presentation:_____

Short Abstract:

Special Technology Requests:_____

Print Request (This is for presentation handouts, not posters. If printing is needed, documents must be sent to Kate Crane by Thursday, May 16, 2013).

◯ Yes
◯ No

Meals and Event Options

Do you have special dietary needs?
☐ Diabetic
☐ Vegan
☐ Vegetarian
☐ Other:_____

If you are vegan or vegetarian, please give us some suggestions for protein dishes that you'd like to see at May Seminar.

Will you be attending the Farewell Dinner on Saturday, June 2nd?
◯ Yes
◯ No

Will you be bringing a guest to the Farewell Dinner? (Contact Kate Crane for more details)
◯ Yes
◯ No

Will you require a vegetarian option for the Farewell Dinner?
◯ Yes
◯ No

FIGURE 12.7   May seminar event PDF fillable form page 2. (Courtesy of Dr. Kate Crane.)

potential issues with programs (such as opening files, saving information, and sending the form back).

In retrospect, it would have been more useful to ask users to provide screen captures and think-aloud protocol audio of their interactions with the sample form so the evaluators would have a better sense of how users completed the form (and differences in steps or order of steps between users). Instead, users reported via email how easy the form was to use and provided suggestions for editing. This was still good information, but finding more information about how users filled out the form,

Additional Options

Do you need a Rec Center Pass?

O   Yes

O   No

If the university requires payment for parking, do you want a parking tag?

O   Regular Permit

O   Handicap Parking Permit

O   No Parking Permit

FIGURE 12.8   May seminar event PDF fillable form page 3. (Courtesy of Dr. Kate Crane.)

how long it took them to fill out the form, and how they saved the form may have helped designers streamline the registration process.

12.5.1   ITERATING WITH PRINT

Although UCD principles were applied when creating the seminar registration form, there were still complications with the use of the PDF form. First, the assumption that all users would have access to Adobe Acrobat Pro to save the form as well as knowledge about how to save and send the form was misplaced. Rather, users filled out the form and selected the print-to-save option when they returned it. This was fine for receiving the information, but the data fields could not be aggregated into one spreadsheet as hoped. Therefore, many of the registration forms' data had to be entered manually into the data spreadsheet. This complication has been seen in other usability studies. In his usability analysis of two writing handbooks, Howard (2008) found that user difficulties "...can all be partially attributed to incomplete understandings of users' goals and task environment. It is obvious that improving an author's understanding of the users' needs is likely to result in more usable information products" (202). In Howard's example, the handbook authors' assumptions about users and their goals for usability evaluation did not match users' preparation for using the product's information. The seminar registration form suffered from similar issues.

Second, although user testing did occur, a convenience sample was used. Users who were known to be cooperative and responsive to the coordinators were targeted for usability feedback. This feedback was used in the final version of the form. However, a few usability problems were not discovered. The first was that new participants in the seminar were unfamiliar with the different presentation types expected during the seminar and needed clarification. For them, and others, this was a complex task because they had to find information about the seminar presentation information in a different document. Additionally, registration began nearly two months before the event began; at that time, participants were not able to provide sufficient descriptions of their work for presentations. Finally, the registration form was only one of two required forms to plan the seminar. Information for housing reservations was collected with a different form. Using multiple forms was confusing and frustrating for students and coordinators. Selecting users with various levels of experience could have helped discover these usability problems.

Creating a registration system for the seminar meant not only going through RABBIT once, but also revisiting the design to alleviate the problems observed in version 1.0. As users and the organization (or event) changes, reevaluation of how the document works for users is needed. Albers (2011) notes that most systems are complex, and the seminar registration form is a fitting example of this. Many different factors affected the usability of the print document: the software used, the information needed to complete the task, and user workarounds that made data collection difficult. However, these findings were incorporated into the following year's registration form, which resulted in a more successful document and process.

Although print documents seem like "traditional" forms of communication and thus may be perceived as easy to use, the examples above show the careful design work needed to develop usable documents. Print is still used, and often preferred, in situations in which information is needed quickly (such as instructions for emergency situations) or when technology inhibits the user from completing a task (like filling out a form). Creating a user-centered model (RABBIT) for users is important for any product—even print.

REFERENCES

Albers, M. "Usability of Complex Information Systems." In *Usability of Complex Information Systems: Evaluation of User Interaction*, edited by M. Albers, and B. Still, 3–16. Boca Raton: CRC Press, Taylor & Francis, 2011.

Durack, K. "Patterns for Success: A Lesson in Usable Design from U.S. Patent Records." *Technical Communication* 44, no. 1 (1997): 37–51.

Howard, T. "Unexpected Complexity in a Traditional Usability Study." *Journal of Usability Studies* 3, no. 4 (2008): 189–205.

Porter, J. "Recovering Delivery for Digital Rhetoric and Human-Computer Interaction." 2008, accessed February 23, 2016. http://daln.osu.edu/bitstream/handle/2374.DALN/54/porter _digitaldelivery.pdf?sequenc=7.

Redish, J. "Technical Communication and Usability: Intertwined Strands and Mutual Influences." *IEEE Transactions on Professional Communication* 53, no. 3 (2010): 191–201.

13 Mobile UCD

The UCD process you use for developing a mobile application should look similar to what you would use to develop other types of products. EyeGuide Focus, to which we have referred to often for examples throughout the book, includes a mobile application designed to run on tablets such as the iPad. EyeGuide developers employed the RABBIT approach for it—researching users and the use situation, balancing feature requirements with client and user needs, and prototyping and testing with real users—just as they did when making the Focus hardware. UCD for mobile application development resembles UCD development of other products.

There are, of course, material challenges that make a mobile application design different from, for example, a desktop application intended to be used by the same population of users. Less screen space means less content; otherwise, users are overwhelmed by too much information on too small of a screen. Images also need to be scaled to fit. Additionally, users don't click with a mouse. They push, pull, and swipe, and because all of this represents different actions, interactive items on the mobile application screen need to be designed (in their shape, size, and results) for fingers to easily use. Remember our "fat fingers" story from earlier in the book.

The mobile application, perhaps more so than any other product, is significantly dependent upon users' contexts. Poor network connectivity causes users to be frustrated with delayed loading of content. The mobile part of the application also implies that users are on the move or, at the very least, using the application while doing other things in dynamic environments. As Nielsen and Budiu (2012) note, because of "small screens [,]…awkward inputs" and "download delays," the UCD mobile application designer must find a way to "limit the number of features to those that matter the most for the mobile-use case" (52).

Ironically, one of the biggest mistakes mobile application developers make, and really the most egregious example of poor UCD, is that many developers choose not to create mobile applications to display content and features on a smart phone or tablet. They dump content on to a small screen that was originally intended to be used on a larger desktop or laptop screen. The result, obviously, is user frustration and failure.

It could be that you don't need to design a mobile application. Everyone, we know, wants to do it because, in some ways, creating a mobile application and putting it on a store for download is the cool thing to do right now. But the first step you should take in designing a mobile application is to find out if your users and also your client really need it. If all of your users or a large percentage of them will use your content and other features exclusively at work, where the primary access tool is a desktop or laptop computer, it may very well be the case that a mobile application is overkill. Should you know, however, from your research of users, as well as an assessment of context and client, that users will primarily be in a mobile environment when they use what you are creating, take the time to create a mobile application.

Other questions to ask, beyond if users need a mobile application, according to Cerejo (2011), include the following:

- Why do they [users] use your site on a mobile device?
- What features are they using?
- What features are crucial for them when mobile?
- What are some sources of frustration?
- What devices do they use to access the mobile web? (sec. 2, par. 2)

13.1   REDUCE, MINIMIZE, SIMPLIFY

A straightforward, best-practice admonition to heed when creating mobile applications is the following: reduce content, minimize input, simplify interface. Your users likely have different goals when using content on a mobile device; it's imperative to know what those are and how they differ from when users access content in different environments, such as on a desktop or laptop computer.

You know the screen is smaller for a mobile application, so reduce image sizes and other gestural interfaces. You can also reduce content. Avoid headaches caused by slow connection speeds. Give users what they need where they need it. Don't make them have to request new content that requires more uploading and downloading. Finally, keep the interface simple and usable for where users will be when they use it.

13.2   CASE STUDY IN MOBILE UCD APPLICATION

13.2.1   NPS Geysers—The First Officially Sanctioned Mobile App for Yellowstone National Park

To aid you in understanding how best to use RABBIT for mobile application development, we turn to an award-winning expert on the subject, Dr. Brett Oppegaard. Over the last few years, Dr. Oppegaard and his team have relied on bedrock UCD principles (see Chapter 3) to develop highly successful mobile applications for use by visitors at national parks in the United States, including Yellowstone National Park. We asked him a number of questions to give you a sense of the steps taken by experienced UCD designers to make effective mobile applications for difficult environments and challenging, diverse user groups.

13.2.2   About Dr. Oppegaard

Brett Oppegaard, PhD (Figure 13.1), an assistant professor at the University of Hawaii, studies ubiquitous computing and mobile media. He was the individual recipient of the regional and national 2012 George and Helen Hartzog Award for his research on mobile app development and media delivery systems within the National Park Service as well as the national 2013 John Wesley Powell Prize winner for outstanding achievement in the field of historical displays. He also teaches communication and

FIGURE 13.1   Photo of Dr. Brett Oppegaard. (Courtesy of Dr. Brett Oppegaard.)

digital media classes stemming from his many years of experience working for daily newspapers, during which he earned several national, regional, and state awards. He was chosen for a National Endowment for the Arts fellowship as a journalist and also has earned National Endowment for the Humanities' grants as a scholar for his innovative mobile media research projects. Those projects include collaborations with America's first national park, Yellowstone, and the National Park Service's Harpers Ferry Center, the Interpretive Design Center of the federal agency. He now works in the school of communications within the college of social sciences at the University of Hawaii's flagship Manoa campus.

13.2.3   Would You Describe the NPS Geyser App and Its Features?

The NPS Geysers app is the first phase of an ongoing project exploring the interactive learning possibilities of ubiquitous computing at one of the world's most dynamic places, Yellowstone National Park. This version 1.0 application (available for free on Android and Apple platforms) was created as a way for our research team to plant a digital flag at the park, to begin to see how visitors to the site might want to use their mobile devices in a myriad of ways, in such a rich natural setting, and to create experimental designs and try them on real audiences in real situations.

One of the most important findings of the first site visit was the disorienting practice in this location of time being conceptualized in a unique way. It was not oriented toward Coordinated Universal Time (UTC) or Greenwich Mean Time (GMT) or even U.S. Mountain Standard Time (MST), the time zone in which this park is physically located. Yellowstone instead operates on "Old Faithful time," as visitors disconnect from the outside world's chronological cues and coordinate their clocks to the predictable eruption patterns of the iconic centerpiece of the Upper Geyser Basin. Old Faithful gushes about every 90 minutes or so, and the persistent question virtually every visitor asks the park staff at some point is, "When is Old Faithful going to erupt?"

As a way to respond to that question and pervasively share such information throughout the area, rangers quickly make their prediction for the next eruption as soon as one ends and then rapidly spread that prediction by radio, by word of mouth, and by foot to various spots on-site through diverse analog media forms stationed in high-traffic areas. Sandwich boards, static clock displays, and signs of various types—including chalkboards, dry-erase boards, and vinyl flip boards—are spread throughout the site and laboriously updated each time the centerpiece geyser blows, to reflect the next eruption prediction and immediately answer the persistent Old Faithful question. Rangers also hustle to spread the word through broad announcements to packs of people and with one-on-one conversations.

All of this seemed to us like a lot of work for information that even more rapidly and widely and more accurately could be spread via the mobile devices that most visitors already were carrying around with them. So one of the first features we created on this mobile app was continually updated prediction pages (for Old Faithful but also the five other predictable geysers in the basin). As the eruption prediction is made and input into the central computer, the mobile app connects with a custom API, pulls this data into its system, and spreads it instantaneously to all app users, including to smartphones, tablets, and even other types of smart devices like watches (Figure 13.2). Such a service appears to have great potential for improving communication efficiency and effectiveness at the site.

Yet off site, we also have found a sizable audience of people interested in knowing when the geysers will erupt. By including in the app the park's webcam view (Figure 13.3), which is one of the most visited pages in the National Park Service web domain (with 3.7 million unique page views in 2015), we have enlarged the outreach to viewers worldwide. In the process, we have made watching the webcam more convenient by giving access to mobile app users and aligning this activity with users' time zones—putting both the alert and the webcam view into this single program.

At the same time, we also brought the park's social media feeds into the app as well (Figure 13.4), allowing users to check Yellowstone's Twitter, YouTube, Flickr, Facebook, and Instagram channels with slick swiping gestures for something to do in-between eruption windows.

The app has a dynamic map available through it; a process for taking a unique digital "postcard"; and overview text, photos, and videos of each predictable geyser (Figure 13.5). Future iterations are expected to experiment with the inclusion of interactive narratives about the history and culture of the site.

FIGURE 13.2   Grand geyser eruption prediction screen. (Courtesy of Dr. Brett Oppegaard.)

13.2.4   WHO WERE THE TARGET USERS?

At first, we envisioned the primary app user as anyone visiting the site with a mobile device in tow. But as we gradually refined our ideas about the audience and potential audience—talking to users, observing users, and tracking analytics—we could see the potential for both an on-site audience and an off-site audience.

About 3 million people visit the site each year, and the two biggest demographic groups were those 15 and younger (23%) and those 36 to 50 years old (26%) with almost half of the travelers coming in groups of four or more. This data meant to us that hundreds of thousands of middle-aged parents come to Yellowstone each

FIGURE 13.3   Yellowstone National Park Old Faithful web cam on mobile app. (Courtesy of Dr. Brett Oppegaard.)

year annually with hundreds of thousands of their children. That understanding has inspired us in future research and development to explore the potential mobile devices have to prompt multigenerational discussions at the intersection of physical place and digital media, particularly when prompted by the incorporation of inter-active narratives.

We also would like to experiment more with the massive—but less defined—web audience of Yellowstone enthusiasts, wanting to determine more specifically who composes this group and how they use the webcam in the mobile app to engage with informal science learning.

FIGURE 13.4   Yellowstone social media on mobile app. (Courtesy of Dr. Brett Oppegaard.)

13.2.5   WHERE IS IT USED?

At this point, the app is primarily used in Yellowstone National Park's Upper Geyser Basin, where it connects to the needs of site visitors by orienting them to Old Faithful Time and providing helpful contextual information. Our second main audience is the dispersed Yellowstone patron, maybe a person planning a trip to the park soon or one who had a great time at the site and wants to keep connected to it in some way afterward. The first audience primarily uses the app locally and then loses interest when Old Faithful Time is no longer a central force in their lives. Yet some of those people also want to keep reliving those moments, and the app provides them with different ways to remain a part of the network, such as getting smartphone alerts

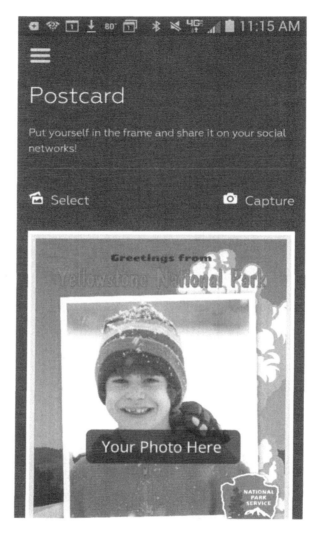

FIGURE 13.5 Digital postcard feature on mobile app. (Courtesy of Dr. Brett Oppegaard.)

about the Old Faithful eruption window opening, providing easy access to the web-cam perspective for witnessing the spectacles remotely, and facilitating easy social media connections.

13.2.6 What User Research Did You Carry Out to Understand Their Needs and Wants?

The design discussions started with brainstorming sessions among researchers, Yellowstone staff, and National Park Service leaders about the potential for mobile technologies at the site, plus anecdotal descriptions of what users typically did at this place, what they wanted to do, and how they wanted to do it, combined with management desires to include safety messages, contact information, and other sorts of service information. These discussions generated several key concepts for us to consider and explore but also a perilous "Swiss Army knife" of possibilities, in other words, a long list of ideas that could be pursued in which the mobile app being proposed could do this and that and this other thing as an almost limitless list of potential paths of inquiry.

Despite the lengthy discussions about these issues, though, the most critical moments in this app development actually emerged throughout the week in June 2014 when I focused on user experience issues at the Old Faithful site with National Park Service staff. That was when all of the general talk we had about Old Faithful Time crystallized for our research team as we watched with amazement at the many ways in which visitors sought the eruption time as an orientation point to the mesmerizing virtues of the place while the staff scurried around behind the scenes endlessly trying to update numerous information stations every time a geyser blew, a pressure especially persistent in relation to Old Faithful.

National Park Service properties have very restrictive and highly managed procedures about gathering data directly from visitors, allowing only small-sample pilot tests and observations. We could contact no more than nine visitors for each prototype test. Yet we did not need more than that in early stages to see obvious patterns among the patrons. To begin with, when Old Faithful erupted, we looked at the crowd—not the geyser—and could easily see that virtually every person witnessing the spectacle not only had a mobile device but also was eager to pull it out to take a photo of the eruption (Figures 13.6 and 13.7).

We instead took photos of the crowd and counted the people with mobile devices versus the ones without. From that exercise, we felt confident that mobile technologies were ubiquitous among users at the site; they wanted to use them to make media about their experience, and one of the key media-making events was the Old Faithful

FIGURE 13.6   People using mobile app at Yellowstone's Old Faithful geyser. (Photo by David Restivo, courtesy of the National Park Service.)

FIGURE 13.7   People using mobile app at Yellowstone's Old Faithful geyser. (Photo by David Restivo, courtesy of the National Park Service.)

eruption. So we built the digital "postcard" activity, in which users could take photos of the eruption, framed with branded park content (Figure 13.5).

We also observed and listened to the incessant questioning of staff about when the next Old Faithful eruption would be, despite the numerous analog signs posted in many places throughout the various visitor centers and pathways. Most frequently, visitors just wanted to know when the next eruption window would start, rarely any deeper details about geothermal science or activity, and each eruption ending spurred a frantic rush around the site of staff members updating signs and announcing the next window. This observation led us to focus the interface design tightly on the No. 1 question people asked, "When is Old Faithful going to erupt?" So when the app is opened, the first piece of information the user encounters is the answer to that question in a bold and clear design (Figure 13.8).

Our next research phase will be focused even more on discovering details about visitors, their current activities, and what they might want to do. For example, considering the demographics of the site and our observations of family members traveling together and walking together and experiencing the site together but not necessarily doing much talking and learning together, we think there is potential for further inquiry into the family-oriented mobile design dynamics of children (15 and younger) traveling with their parents and grandparents. We saw many children in that age range disengaging with their surroundings to look at social media on their phones or play mobile games. We think apps with engaging content and the right kinds of mobile interface design and activity prompts could instead get the devices used more as a social facilitator between generations.

So some of our next research phases on this project will be to investigate interactive storytelling features, family-oriented interactions, and mobile technology affordances that matter the most to that user group. That process will include conducting more contextual inquiries of user activities, more user shadowing, and deeper collection of user-profile data as a way to create personas, user journey

FIGURE 13.8 Old Faithful geyser mobile app main screen figure. (Courtesy of Dr. Brett Oppegaard.)

maps, storyboards, prototypes, and field experiments, testing representative users in authentic situations.

13.2.7 WHAT WORK DID YOU DO TO ASSESS THE SITUATION(S) OF USE?

Besides extensive observational work, like the examples described above, we also created prototypes of various fidelities—from paper to simple beta apps—that were used on small samples of representative users in real contexts to gather critical understandings about ways in which visitors might want to interact with information at Yellowstone. While our focus was on generating insights for mobile app development, we also took into consideration the greater media ecology and experimented with other media forms, such as the existing signs, in terms of interviewing visitors about their uses of the analog material and their preferences for receiving such information, including testing their interest levels in communication through mobile devices and trying to tease out what that mobile form might look like in their projections of it.

To consider other possible courses of action, we experimented with digitally projecting the information on the walls of the main visitor center. While we determined no single solution could solve all communication issues we witnessed, we also saw several clear ways in which mobile technologies could improve the media ecology of the site and allow for potential removal of some of the clutter and aesthetic inconsistencies of other media technologies, such as hand written whiteboard scrawls.

13.2.8 What Challenges Did You Face in Balancing User Needs and Wants with Environmental, Material, and Other Factors?

We originally had many grand ideas about what we possibly could do with mobile technologies at the park, such as tracking and showing bear activities in real time, creating a massive multiplayer contest encouraging citizens to compete with each other about eruption predictions, and developing elaborate interactive narratives. But users, without much doubt, were most concerned at this point about getting the answer to the single overriding question, "When is Old Faithful going to erupt?" Because mobile technologies could help significantly in solving that quest for visitors, improving both efficiency and effectiveness of park–visitor communication, we felt compelled to start there, even though as researchers we had many more intriguing paths pulling at us. Yet we also thought if we solved that core concern and built an audience from that effort, an audience who trusted us and our work, they later would come back to us when we tried more ambitious and creative endeavors.

In addition, developing the more complex activities imagined, as a first major step, inherently would be more costly, more time-consuming, and more risky than addressing the Old Faithful Time issue. Furthermore, as we worked at the site with prototypes, we encountered significant concerns about the reliability and speed of the network in the area as well as the strain on Internet bandwidth at peak times (such as when Old Faithful was erupting and hundreds of people were trying to put photos or videos on social media channels at once). With all of those considerations, we decided to address the practical concerns first and build in phases toward more complicated issues, such as exploring what family members might want to do with their mobile devices that would involve learning and social interaction with family and friends but also, possibly, other people on-site or even just folks interested in the park but watching from afar.

13.2.9 How Many Prototypes Did You Create and of What Levels of Fidelity?

We created dozens of digital prototypes throughout the process, including many that never left the alpha stage of testing, but on-site, we experimented with low-fidelity paper prototypes as well as high-fidelity test apps, applied to single-issue niche design concerns, such as questioning whether users want to swipe or scroll or click through segments of the design and also large-picture activities, such as asking representative users to take customized photos of family members in front of Old Faithful when it was erupting.

13.2.10 What Was the Value for You of Testing Users in Realistic Situations Using Prototypes of Your Product?

Giving the prototypes to representative users in real contexts generated a bonanza of insights to us, both small and large. The designer's perspective can be blind to many practical concerns, such as an issue we had with our high-fidelity prototype in which users did not often recognize that more information could be found about each geyser in the app by simply scrolling down or swiping left or right. When we quickly added a tutorial overlay to the interface, many of those navigation issues suddenly disappeared. I think that because we were so invested and interested in this project, we would naturally want to turn on the app and play around with it and see what it could do. But the user, in this case, might not be as motivated (and most likely wouldn't be). So this person might want to simply turn the app on and find out the next time Old Faithful is likely to erupt. The user might not want to take a photo with a fancy digital frame, or look at the webcam, or explore the park's social media. For this person, the app might only be useful for finding that eruption time, and as researchers, we adapted our use case expectations to more realistic probable user engagement levels. When conceptualizing the app from that base user's perspective, we aimed to satisfy the foundational needs first, and from that initial point of satisfaction, we are hoping that other interesting experiences can grow.

13.2.11 What Best Practices Can You Suggest to Readers with Regard to Building Mobile Applications?

Assume almost nothing about your users, and instead spend as much time as possible at the start of the project learning about who they are, what they might want to do, with whom, and in what ways and where, preferably in situ. Then, look for problems in those experiences and try to solve them. The competition for attention and activity in contemporary society is so fierce that those who want to research and develop mobile apps as a way to build communal knowledge (both scholarly and practical) are better off, in my mind, trying to solve clear user issues, like determining when Old Faithful's eruption window opens next, than expecting users to have an inherent motivation to follow wherever a scholar might want to take them.

From my perspective, when I have started design conceptualization at the juncture of the user and the problem that the user is encountering, within a particular media ecology, and then considered how mobile technologies might make that situation better, I have had much more generative and productive experiences than when I have followed my whims, wherever they were leading me, and expected users to line up and follow my meandering twists and turns as a return on my efforts and investments. While such expeditions can pay off and break ground, they also can lead one metaphorically into the woods while other scholars are busy solving the problems of real people in real contexts in real places.

REFERENCES

Cerejo. L. "A User-Centered Approach to Web Design for Mobile Devices." *Smashing Magazine*. May 2, 2011, accessed February 25, 2016. https://www.smashingmagazine.com/2011/05/a-user-centered-approach-to-web-design-for-mobile-devices/.
Nielsen J., and R. Budiu. *Mobile Usability*. Pearson Education, 2012. Kindle Edition.

14 UCD Teams

At a company Brian worked for some years ago, the technology team was in search of a new server administrator. The hiring committee, including Brian, had narrowed down a larger list of applicants to three finalists. The last of these finalists to be interviewed had previously done some contract work for the company, and he also knew the technology manager. You could say that, given this, he had an inside edge to getting the position.

During the course of the interview his answers to questions revealed a great deal of competency. He showed himself to be experienced and creative, and all of that, in theory, would make him an ideal choice to administer what was a complicated mix of different servers and applications along with an accompanying computer security infrastructure. If only he hadn't answered the final question of the interview as he did, he surely would have gotten the job.

It was a simple question, asked by the human resources (HR) manager of every one of the three finalists: "How do you see yourself as part of a team?"

"That's a great question," he said. He had trained himself well for the interview. Complimenting questioners is always a good tactic.

"Well, I see myself a lot like Clint Eastwood. When there's a problem, I ride into town on my horse, I see the bad guy, I shoot him, and I ride away into the sunset. No more bad guy, no more problem. I solve problems."

There were a few chuckles and then closing remarks and the interview ended. As the committee discussed the finalists, an interesting conclusion emerged. Everyone on the committee agreed the last candidate had the best skills and experience for the job, but everyone also agreed they didn't want to hire him because they didn't see themselves enjoying working with him. They saw his attitude as not problem solving but in fact problem creating.

"We're collaborative here," one said.

"We work together, we have each other's backs, and we communicate with each other so that we know what's going on," another said.

"We've got enough macho men running around. No need for another one. And didn't Clint Eastwood sometimes play a bad guy? We don't need our town shot up every day. It's hard enough as it is to do the job."

Although the best candidate on paper had worked as a contractor at the company before, what he didn't grasp during that time was that a gunslinger, individualistic attitude toward teamwork and problem solving was antithetical to how the technology team saw itself or wanted to operate. In the end, the second ranked of the three candidates, a little less experienced but seemingly more amenable to a collaborative way of doing things, was hired. Brian actually ran into him a few years ago, and he was still on the job, now in a more senior position.

Every team or group of people working together is different and is heavily influenced by several factors:

- Organizational culture
- Individual personalities and experiences
- Workplace setting (virtual vs. on-site)
- The nature of the product, processes, or services the team is assigned to create or manage

You may find yourself in situations in which you are the only team member and do not have to work directly with others to manage your activities. Still, it is almost always the case that even the individual designer is answerable to someone, whether that be a manager directing that designer's work at some level above in an organization or a client with whom the designer is engaged as an independent contractor or consultant to produce a deliverable. The point is that in UCD, we are, as individuals or members of an obvious team, interconnected to others. At the very least, we are engaged with users as much as possible.

Therefore, how a team works, how you can contribute to or manage such a team in a UCD environment, and what the characteristics of an effective team member are all are important questions for you to consider. Because of this, the answers to them are the focus of this chapter. It is to your benefit to understand teamwork as much as it is to understand the other important aspects of fundamental UCD product development.

14.1 TYPES OF TEAMS

To begin, for our purposes, we will define a team as a group of people tasked with developing a UCD product. Obviously, there are other kinds of teams, and if your design team is in a larger organization, it could be influenced by other teams; you may even find yourself on another type of team but also assigned to the design team. Still, we will go too far adrift of our aims of discussing UCD if we spend too much time looking at team parameters and dynamics that are beyond the immediate concern of building UCD products. When we discuss teams here, we mean UCD-related teams.

Given this definition, we'll note that there are different types of teams and, correspondingly, different ways you can contribute to or participate on such teams. Additionally, dependent on the team, who is part of it can vary. We'll break down a few of the more common team types, explaining briefly how each one works and what factors influence those workings.

14.1.1 Dedicated UCD Team, Dedicated UCD Organization

You will probably consider the dedicated UCD team, as part of a UCD-driven organization, the ideal team. In this situation, the organization at all levels puts user needs and wants first. In all likelihood, the products it makes and sells are not only created using a UCD methodology, but they are also marketed and sold to targeted users as being about and for them. UCD is a way of doing business for such organizations.

Such organizations may have one team that does all design and support or if it is larger with more mature product lines or if it offers design services to other companies, it may have multiple teams that contain different roles all working together to create UCD products. In the latter, you may be a designer assigned to just one product or you may work on different teams designing different products. Should the organization have just one team, you will probably work on multiple design products, creating new ones or maintaining existing ones already on the market.

UCD-driven organizations can be hierarchical or top-down in their management strategy, meaning leaders at the top of the administrative structure make decisions that are passed down through the chain for others to manage and carry out. However, in most cases the pure, UCD-dedicated organization is by nature more collaborative and open in how it operates. Employees see themselves as collaborators, all capable of interjecting ideas about the organization's operation, even communicating them directly to senior leaders. Regardless of the management style of the organization, the one thing that separates the UCD-dedicated organization from others at which UCD may occur is that there are no barriers, no lack of understanding, and no resistance to UCD as a design approach. The organization makes its money on UCD, so it is committed in every way to putting users as the focus of design. There may be disagreements about how to do this, about what the right design decisions to carry out based on feedback from users or other types of analyses conducted are, but everyone is committed to UCD. It is how the organization makes money and fulfills its core mission.

To be an effective team member in such a UCD-dedicated organization, you have to be skilled, obviously, at UCD and also possess some kind of design-related abilities. Further, you have to be confident, communicative, and open to giving and taking feedback. As great as it would be to work in an organization that gets and practices UCD as its core mission, the people that surround you know UCD and are experienced, talented, and demanding. You likely aren't going to spend a lot of time learning on the job. You need to be able to contribute effectively to enhance the organization's UCD efforts.

14.1.2 UCD Representative as Part of Dedicated or Ad Hoc Development Team

As a designer, you will probably be involved in a more common situation as part of one of the following: a development team or an ad hoc team.

14.1.2.1 Dedicated Development Team

In this situation, you are the designer tasked with directing UCD efforts because of your knowledge and skill in the approach. The team you are assigned to may develop software, websites, mobile applications, even hardware, but it is comprised of developers, engineers, business analysts, and other roles that may or may not be interested in or have any understanding of UCD.

For example, if you are brought on to be a designer responsible for UCD as a member of an agile team, often called a scrum, you will have to integrate UCD into an already existing agile methodology for development. In some respects, UCD will work well in such an environment, but there will be aspects of how UCD works that agile developers on this team might regard as time-consuming, getting in the way of their ability to develop and release new code as quickly as possible in time frames called sprints.

Unlike the UCD-dedicated team, the development team and, for that matter, any other team tasked with development but using another methodology to work will require that part of your contribution to the team, in addition to being a designer, is to advocate and educate the usefulness of UCD so that it can be better integrated into the team's existing development processes. This may not be easy, and it can cause some friction. If the team culture is more open and collaborative, it will be easier for you. If team members, however, are locked into rigidly defined roles or if the manager of the team has a lot of power to decide how development works and is not in favor of UCD, it will be more difficult for you to assert the value of UCD.

14.1.2.2   Ad Hoc UCD Team

An ad hoc UCD team is one created for a specific purpose, and it is often the case that this purpose is for a project that occurs only once or infrequently. For example, an organization may create a new website or other product intended for users, inside or outside the organization. You may belong to the organization and already do design work for it or you may be brought in as an independent contractor to provide design skills as part of the ad hoc team. Most of the members of such a team are typically pulled from elsewhere in the organization, so their contributions to the ad hoc team are not full time. In many cases, they are also not expert contributions. For example, a representative from accounting may be on the team to speak on behalf of accounting in the design of the organization website although that representative may not care and probably has no knowledge of how to design an effective website.

Interestingly, you can see the results of ad hoc team efforts in the many poorly designed organizational websites still out there. Because ad hoc team members aren't experts and what they know about the company is informed by their own internal perspectives, invariably, the website created isn't built for external users; it reflects internal goals and internal views of the organization. Users are given information about management structure or biographies about leaders when they really want to know how to use the services the organization offers.

Your job as a designer on such an ad hoc team, much like your role on a dedicated development team, is to be a UCD educator and advocate. There may be less resistance to UCD, which is good, but there will be a lack of knowledge about how to do any kind of design, let alone UCD. At some point, on a dedicated development team, you could win over the other team members to UCD, and they would have the skills to implement it as a methodology for current and future products. On the ad hoc team, the skills aren't there, and on top of it, the team will go away once the product is completed. Consequently, you will have to take on a lot of the heavy lifting yourself in implementing UCD and within a smaller window of time. But because you may not be a manager or even an employee of the organization, you need to gather up a strong understanding of the culture of the organization, the decision-making process, and the personalities of the team members and other key stakeholders so that you can construct a plan that aids you in building up support for implementing a UCD approach.

This isn't easy. It's no wonder that Unger and Chandler (2012) assert that one of the most important parts of implementing UCD for an organization involves first examining carefully its history, logistics, and management hierarchy (43–46). You need to get answers to the following questions:

- What past projects were carried out? Were they successful? If not, why did they fail?
- What technology is at your disposal to do your work? Does the organization support remote or virtual work?
- What is the management culture of the organization?

Addressing this last question, Unger and Chandler point out that "power distance within the...company will have an impact on how you successfully navigate the political waters during the project" (45). In other words, if the organization is top-down in how it manages, a few senior leaders making decisions that are distributed down a chain of command to be carried out with no feedback or modification possible, then the entire organization will be hierarchical in nature. Therefore, to introduce change or to get buy-in for UCD, should it not be a currently implemented design methodology, you'll need to either get the senior leaders to advocate for it and force its adoption through the organization or convince managers with control closer to the development process itself, who have backing from senior leaders, to support it. And they may even be reluctant to go against their superiors. In large power distance workplace cultures, there is not much room for flexibility for moving outside of assigned roles to collaborate with others or introduce change unless it is mandated and/or approved by the existing management hierarchy.

14.2 UCD TEAM ROLES

You may occupy two other roles as you participate in UCD: an independent, lone wolf designer and UCD team manager.

14.2.1 Lone Wolf UCD

This is great. You are your own boss, not working for anyone as you create incredible products that effectively serve the needs and wants of users in the places where they will be used. But unless you are making something that just sells itself without the need for marketing, business, legal, and other expertise, you're going to have to create your own ad hoc team to help you make known to the world what you've created. Should you not go that route but instead opt to offer your independent services to outside organizations, you will need to market those services, creating a brand for your lone wolf operation that tells prospective clients you are the best designer out there and that UCD is the best solution for them, in combination with your skills, to get the job done.

What we're basically asserting here is that there's no such thing, in most cases, as a lone wolf, or at least not a successful one. You should be prepared for the politics of teamwork, but you shouldn't eschew the team concept. Even if you are trying to go it on your own, find ways to create a team of support for you. Join professional associations, such as the User Experience Professionals Association (UXPA), The Design Society, or the Association of Computing Machinery (ACM) and its special interest group on the design of communication called SIGDOC. Such groups allow you to talk with others, sharing and getting advice online, on the phone, or even in person at weekly/monthly local meetings or national and international conferences. You will like UCD more—you will even find it more successful for you—if you align yourselves with others who do UCD.

14.2.2   UCD Team Manager

A role you may find yourself in, either because you advance in seniority at an organization or because you are brought in for the explicit purpose of becoming the new design leader, is that of UCD team director. You could be leading a team of fellow designers or a team of designers as well as other technical and development roles (i.e., mechanical engineer, business analyst, marketing representative, content manager) or a team made up of ad hoc personnel (working on the design part-time while holding down other organizational jobs). Whatever situation you're in, as the team leader, you are the one responsible for fostering a productive team culture that allows for the successful delivery of the product.

14.2.3   Team Management Best Practices

You want to put yourself in a position to be effective, and adhering to the following best practices, all or in part, will help.

14.2.3.1   Put It in a Plan

You can't "wing it" when it comes to leading the development of a UCD product. You already have noticed, if you've read this book chronologically, that a lot of research goes into understanding users, their needs and wants, and their situations of use. Added to this quantity of necessary information that must be gathered and documented is also the practical information of running a team, things such as deadlines, team roles, and budgets that must also be planned out and managed.

You don't have to be a certified project manager, but you do have to put in place a plan for how you're going to get from where you are at the beginning of a product's development to the end when it goes live and becomes available for everyday use by its users. You might ask yourself the following questions:

- What resources do we need to complete the product design? People? Tools? Money?
- Who will be responsible for what?
- What are the deliverables?

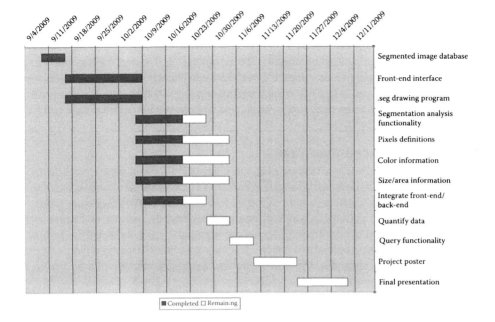

FIGURE 14.1 Gantt chart showing a project's planned deliverables on a timeline ("CSE 379 Gantt Chart for the week of 10/21," Alex Thompson and Derrek Harrison, public domain).

- What are the key dates or milestones for these deliverables?
- How will we research, design, and test?

All of these are significant questions your plan should account for, and you can make use of a variety of tools to manage it. Gantt charts, for example, have long been used to manage a project's life cycle. Figure 14.1 shows a Gantt chart constructed by a design team, BSCM, to satisfy a proposed content management system development project. You will see a list of all key deliverables on the right of the chart, then a graphical representation of when each deliverable will be achieved within the context of the larger project's planned timeline.

Planning a product's development will prevent mistakes. Be prepared to change as you need, but put that flexibility into the plan itself.

14.2.3.2 Put On a Show

Don't let your team work in isolation without anyone knowing what you're doing. It may be that what you're creating is considered intellectual property and as such can't be revealed to just anyone until it is protected. Find out who can see it, and let them see it as you work on it. Even if you can't show off what you're creating, let everyone know what your team is doing. Invite them to open houses, publicize the work you're doing, and even think about creating in a space open for eavesdropping so that the wider organization knows what you're doing.

Invite controlled participation; doing so creates transparency, excitement, and support for your work. EyeGuide was at a conference a few years ago, and one of the team's developers wore around a headset prototype as he visited different vendor booths. Multiple people came up and asked him about it, wondered what it was for, and also offered suggestions for how it could be used. Later, more than one person tweeted the experience, and after that, some signed up to follow the company on Facebook. Put on a show every way you can so that people know what your team is doing.

14.2.3.3 Put Users on Your Team

Nothing better reminds you of user needs and wants than users sitting right next to you, helping, in fact, design the product. You have to be careful to listen only to the users on your team because at some point they become, as part of the team, participants behind the curtain, and so they do not see the world the same way as their user counterparts in the wild, real world. You can mitigate this cynicism brought on from experience by rotating users who occupy roles on your team. But even if you don't, having users work with you offers an immediate, tangible reminder of why you're doing what you're doing.

14.2.3.4 Put the Right Skills on Your Team

Winter (2015) argues that in addition to a team leader, a good UCD team should have the following members:

- *UX designer*: Responsible for researching and designing the overall user experience, including conducting user testing, prototyping, doing field research, and combing through analytics.
- *UI designer*: Responsible for making sure users can interact with and understand the product as intuitively as possible through the design and layout of visual elements.
- *Content strategist*: Responsible for making sure the words users read are clear, easy to understand, and in line with your brand; content is just as important as the visual design.
- *Developer*: Responsible for bringing all your team's great ideas to life with code.

All of these positions, of course, satisfy most requirements for the actual building of the product, but we assert that you should try to bring on others who may not be skilled like those occupying these positions but nevertheless may add value to your team, depending on the type of organization to which you belong. We already mentioned user representatives and their value on your team, but, if you can, you should include, even if they aren't active in the process, key stakeholders from around the organization who buy into UCD and could provide you political support as you build out the product and later as you attempt to get buy-in for it after it is done. Business and marketing analysts, customer service experts, others with particular knowledge strengths, and even those with a lot of organizational history or significant interaction with your targeted users can be helpful additions. Obviously, should you be tasked with designing hardware, engineers for electrical and plastics components of the product will also be required.

A larger team can be unwieldy, and any time you introduce a new member to it, you have to consider the impact on the overall team's culture. Think about, in ideal circumstances, that the more active part of the creation team has four to six full-time members, but that the larger team, with people who play part-time, non-essential roles, can be twice that size and still be effective. The more people think they are part of the effort, the more likely they are to want to support that effort to be successful.

14.2.3.5 Put In Time to Practice Sharing, Experimenting, and Solving Together

You should promote hands-on activities that prevent your team members from hiding inside of their roles or looking at the product from behind a narrow disciplinary prism. There is a movement now in business and technology called DevOps, which essentially advocates for developers and operations or systems administrators to merge their thinking together. Instead of thinking about what they do as being opposite to what the other does, developers and operations personnel are encouraged, through technology as well as management practice, to engage with each other and see what they do as being the same thing: the delivery of quality products for users.

Work together on team-building activities that require all of the team, regardless of their designated roles or job titles, to experiment with problem solving and other work that makes them get out of their comfort zones and makes them work with and come to rely on each other in open communication to create solutions to problems. Two excellent resources for team-building activities are the United States Agency for International Development's (USAID) Team Building Module Facilitator's Guide (2012) and J2N Global's Journey to Newland program. USAID's guide provides detailed team-building exercises and discussions on five key areas: characteristics of effective teams, building and maintaining teams, collaborative communication, managing conflict, and monitoring team development.

J2N Global's *Journey to Newland* book (Pool et al. 2007) and corresponding workshop program has interactive products (i.e., workbooks, videos, games) that engage team members on a simulated "journey" on which they encounter unique challenges and must work together in innovative ways to reach specific goals. Beyond helping team members get out of their comfort zones, communicate, and work together to solve problems, *Journey to Newland* helps create a shared vision and set of expectations about what it means to be part of a project team; team members see the bigger project picture beyond their own tasks and realize the value of new approaches with different people.

You can do team-building work as part of team retreats or other teamwork exercises. Although these activities are not related directly to the development of a product, they will push your team to function less like a group of individuals and more like a unified collective focused on working together.

14.2.3.6 Put It on the Board

There are voices out there that argue against whiteboards, asserting that they are often excuses to conduct unproductive groupthink exercises that only confirm the same ideas and don't promulgate new solutions. Maybe. But we like whiteboards. At EyeGuide, every wall is a whiteboard. It encourages team members to work on something in an open way, and, given its size, it also encourages others to step up and contribute.

We want to get rid of role isolation. Avoid cubicles, avoid any opportunity to hide and not engage. Yes, designers need some space to think and craft, but the most important meeting space for your team should be open and should enable communicating, planning, and designing among everyone. A whiteboard does that for sure. You can use it to plan, design, and also test and, after that, evaluate test results and then plan again. It can serve as an artifact to be referenced not just for a current project but also future ones. Other tools may function in the same way such as flipcharts or standard chalkboards. You might even consider creating a team wiki or open chat space in which team members can easily add and respond to ideas. If your team is virtually distributed, take advantage of screen-sharing and recording options. Many browsers, like Microsoft Edge, allow for real-time annotation and capturing; this way, everyone on the team can engage with the "big picture" of the project in the same space. The flexibility of these tools, much like lo-fi prototypes, invites contribution and collaboration; anything can be easily erased, revised, and revisited. Whatever tools you choose, they should help you take ideas out of heads and put them out in the open for discussion.

14.2.3.7 Put Everything on the Table and Talk about It Openly

Along the lines of promoting a better team through team-building activities, you must create an environment as a UCD team leader that is free for anyone to communicate their feelings and thoughts regarding the product and its development process. All team members have to own what they say and shouldn't be allowed to pass the blame on to someone else. At the same time, you must encourage them that there is no blame and no such thing as a stupid idea or comment. Your team, including you, should put all feelings aside except those of the users. Open, honest communication, team members talking to each other and also talking to you and in that effort speaking truth to power, is absolutely necessary.

Borrow from agile development methodology to set up daily, if not weekly, standups during which team members have to talk about what they are working on and get feedback from others. Criticism is fine, but for anything negatively suggested, the same person offering the negative critique has to offer positive solutions.

Lead by example. Show your team that you can take feedback, so they know how to do it constructively, just as they know also how to give feedback in ways that make the teamwork better together. Dickerson (2013) writes, "You need to be there for the team, and a lot of that involves listening to a lot of complaints and gossip. Be encouraging and offer support. Be ready to intervene if your team members need help. And try not to gossip back if you can help it. That feeds the beast and can result in demotivated team members" (par. 6).

14.2.3.8 Put Yourself in the Right Position to Advocate for Your Team

Your job is to lead your team to develop the best possible product, and doing that means representing their interests and the work they do to the larger organization you are contracted with or to which you belong. One of Brian's former baseball coaches once told him and his teammates something that still sticks with him: "When the game is won, we won it. But when the game is lost, I lose. We win as a team. I lose. The team always wins."

Brian's coach wanted to pass on the message that success comes through team effort. At the same time, he wanted also to offer up a message on leadership. A good leader motivates a team to win but also is willing to own defeat, to take the brunt of criticism or the negative fallout from failure so that the team isn't damaged. If the team sees the leader sacrificing selflessly like this, if the team sees the leader so willing to advocate on behalf of the team that the leader takes all of the blame when failure occurs, the team will have more trust in the leader and more motivation to work together to be successful. Dickerson (2013) again has helpful comments here: "If a design project goes well, give full credit to the people who did it. And if things go badly, don't throw anyone under the bus. As the manager, the buck stops with you, and you need to take responsibility. UX professionals tend to be a perceptive lot, so members of your team will notice this and appreciate you for it" (par. 11).

14.2.3.9 Put Away Old Ways of Thinking about How to Work

As much as face-to-face (f2f), open communication among team members and users is the best approach for developing user-centered products, it is no longer practical in many situations. Organizations have employees or contractor labor in different parts of the country or world. The management of virtual team members who supplement an on-site f2f team as well as the management of an entirely virtual team (an increasingly common occurrence), presents challenges for UCD:

- Creating an open, collaborative workspace
- Interfacing regularly with users to get their feedback
- Managing project timelines with team members potentially spread around the world and living in various time zones

All of these and more are just harder to accomplish effectively when some or all of the team works virtually. As a remedy, Yiu (2013) urges, in part, that you have at least a kickoff meeting for your virtual team that is f2f. It might be expensive, but budget for it if possible. You can use the time brainstorming "about writing scenarios," Yiu writes, "[and] design process, running focus groups, or setting survey questions" (2335).

Watkins (2013) also argues for establishing regular meetings, making use of the best possible communication technologies with video and audio capability, and enforcing a consistent shared language: "When teams work on tasks involving more ambiguity, for example generating ideas or solving problems, the potential for divergent interpretations is a real danger...Take the time to explicitly negotiate agreement on shared interpretations of important words and phrases, for example, when we say 'yes,' we mean...and when we say 'no,' we mean...and post this in the shared workspace" (sec. 6, par. 1).

EyeGuide, as previously noted, employed a number of vendors in India and China who worked with the on-site EyeGuide team to develop Focus hardware and software. At the outset of their virtual work together, the gap in working time and poor communication connectivity caused problems. When, for example, the Indian electrical engineers were working during their regular hours, EyeGuide counterparts were sleeping. When communication was synchronous, it was often over Skype, and connections dropped constantly.

Language and cultural communication differences also presented challenges. For example, the Indian contractors preferred to include new information in the replies of existing—often lengthy—email chains. EyeGuide developers, acting on their own cultural communication preferences about email, didn't always see this information in time (if at all) to make appropriate changes. Cultural expectations about being straightforward and meeting client requests also complicated things. EyeGuide developers said, without filter, what they thought and wanted. The Indian contractors, in their reluctance to say "no" to a client, often gave vague answers or no answer at all. Communication preferences, such as levels of directness, formality, and tone often vary among cultures (Stewart and Bennett 1991; Hall and Hall 1990; Yum 1997; Warren, 2004; Dautermann, 2005; Wang, 2008); this was a challenge for both EyeGuide and Indian team members.

Problems and corresponding project delays as well as frustration at EyeGuide increased to such a level that some in management considered seeking out alternative vendor support. But the EyeGuide team leader jumpstarted a new approach to managing the virtual team by paying for the Indian contractor's project manager to come to the United States and work on-site for a week with the EyeGuide team. This fostered trust, and in the process, a schedule for meetings and a system for managing project details were put in place. Also, a more reliable online communication tool was experimented with and adopted. Although some issues continued, primarily related to linguistic differences, the virtual work after the visit became more effective.

Virtual team management is difficult. As Watkins notes, "With virtual teams… coordination is inherently more of a challenge because people are not co-located. So it's important to focus more attention on the details of task design and the processes that will be used to complete them" (sec. 2, par. 1). It requires more effort and planning on your part to make sure team members, working at a distance, are sharing openly, but it can be done.

14.3   EFFECTIVE TEAM MEMBER QUALITIES

UCD is a mixture of business and creation. As the team leader or contributor, you are responsible for identifying which traits and abilities will work best for your team and your project. Your team members should possess some mastery of UCD skills. But being an effective contributor on a UCD team takes more than flexing your individual skills. Beyond UCD prowess, Driskell et al. (2006) consider the following team member traits conducive to team performance: emotional stability, extraversion, openness, agreeableness, and contentiousness. Although each of these traits can have different value under different circumstances, don't limit yourself to this— or any other—list when picking your team.

When you determine what traits you are seeking, cultivate a team culture that incentivizes these traits. If you desire openness, practice an open-door policy. If you want confident, secure team members, be explicit about the positive aspects of their work and demonstrate your trust in them as you assign them new tasks. It may be (which is highly likely) that not every single team member will share the same traits. Luckily, a team full of only agreeable extroverts or any other homogenous characteristics isn't the answer. Above all else, look for team members who believe in the UCD philosophy, care about their work, and are willing to learn and develop themselves into dependable, committed team members.

REFERENCES

Dautermann, J. "Teaching Business and Technical Writing in China: Confronting Assumptions and Practices at Home and Abroad." *Technical Communication Quarterly* 14, no. 2 (2005): 141–159.

Dickerson, J. "Tried and True Tips for Getting the Most Out of Your User-Centered Design Team." *UX Magazine*. March 13, 2013, accessed February 23, 2016. https://uxmag.com /articles/how-to-effectively-manage-a-ux-design-team.

Driskell, J., G. Goodwin, E. Salas, and P. O'Shea. "What Makes a Good Team Player? Personality and Team Effectiveness." *Group Dynamics: Theory, Research, and Practice* 10, no. 4 (2006): 249–271.

Hall, E., and M. Hall. *Understanding Cultural Differences*. Yarmouth: Intercultural Press, 1990.

Pool, B., K. Gray, and G. Gray. *Journey to Newland: A Road Map for Transformational Change*. Hoboken: John Wiley & Sons, 2007.

Stewart, E., and M. Bennett. *American Cultural Patterns: A Cross-Cultural Perspective*. Yarmouth: Intercultural Press, 1991.

Unger, R., and C. Chandler. *A Project Guide to UX Design: For User Experience Designers in the Field or in the Making*. Berkeley: New Riders Publishing, 2012.

United States Agency for International Development. "Team Building Module Facilitator's Guide." *United States Agency for International Development*. September 2012, accessed February 24, 2016. https://www.usaid.gov/sites/default/files/documents/1864/Team -Building-Module-Facilitators-Guide.pdf.

Wang, J. "Toward A Critical Perspective of Culture: Contrast or Compare Rhetorics." *Journal of Technical Writing and Communication* 38, no. 2 (2008): 133–148.

Warren, T. "Increasing User Acceptance of Technical Information in Cross-Cultural Communication." *Journal of Technical Writing and Communication* 34, no. 4 (2004): 249–264.

Watkins, M. "Making Virtual Teams Work: Ten Basic Principles." *Harvard Business Review*. June 27, 2013, accessed February 23, 2016. https://hbr.org/2013/06/making -virtual-teams-work-ten.

Winter, J. "4 Steps to Build an Awesome UX Team." *User Testing Blog*. August 18, 2015, accessed February 23, 2016. https://www.usertesting.com/blog/2015/08/18/build-ux-team/.

Yiu, Charles. "UX Design with International Teams: Challenges and Best Practices." Proceedings of the Association for Computing Machinery Conference on Human-Computer Interaction, Extended Abstracts, Paris, France, April 27–May 2, 2013: 2333–2336.

Yum, J. "The Impact of Confucianism on Interpersonal Relationships and Communication Patterns in East Asia." In *Intercultural Communication*, edited by L. Samover and R. Porter, 78–88. Belmont: Wadsworth, 1997.

15 UCD Tools and Technologies

In the earliest days of the WWW, when websites were beginning to go online with greater frequency, and newer, more capable browsers allowed developers to post images along with text, even interactive forms, the tools available for content creation were, to put it nicely, somewhat crude compared to what can be acquired and used today. HTML code editing, for example, was routinely done in Notepad, a text editor. There was no code validation, no what you see is what you get (WYSIWYG) view. You committed your code to Notepad then used file transfer protocol (FTP) to upload it directly to the web server, in painfully slow kilobits, not megabits, per second. If you were lucky, everything was right. If not, you had to repeat the entire process all over again until you got it right.

Still, we hear every day what some of the more experienced developers say when asked, "So what tools do I need to create an effective website, for example?"

"Well," they say, "start with the one that sits on top of your neck. Use your ears to listen to users and your brain to envision solutions for them, and the rest is easy."

Of course, these salty, old-timers are right. The best tools at a developer's disposal are ears to listen to users and a brain to comprehend and complete actions. It's also true that pencil and paper can be enough to create a decent prototype to get feedback from users. In fact, low-fidelity prototyping like this, as we've already noted, is a democratic approach to development that you should consider strongly as your first, if not only, step. Experts aren't required to make something better; money isn't required, beyond the pencil and paper, to enable experts to create something better; and you don't have to wait until the experts work their magic, which means users can be part of the process as soon as possible.

Nevertheless, in a book like this, trumpeting itself to be a comprehensive overview of UCD, we would be remiss if we didn't offer you some options for tools you can use to carry out effective UCD. So in what is the final chapter of the book, you can see a selection of tools, with screenshots in some cases to demonstrate their functionality, that you can employ throughout development, beginning with user research and ending with user testing and its accompanying data gathering and analysis.

For each tool, we tell you about features and pricing. If you like a tool you see, do a simple Google search to find where it currently resides (web addresses change frequently).

15.1   USER RESEARCH TOOLS

15.1.1   Pendo

If you want to understand how users use your product, such as your website, Pendo provides real-time data, presented in a variety of formats, on who exactly is using your product, when they are using it, and how they are using it. If you are creating a brand new product, Pendo might not be useful, but if you are revising a current product, or if you have just deployed your new product so that users can interact with it, Pendo can provide a lot of data (Figure 15.1), akin to Google Analytics, for understanding what users are doing. You can even use Pendo to target these users as they use the product with what is called in-app messaging.

When we checked, prices started at $99 a month, so this is something you would budget to give you a lot of good data to inform redesign or to keep abreast of user feedback after you've gone live so you can tweak your new product to keep it effective.

15.1.2   SurveyMonkey

You likely have answered (and maybe already created) a survey thanks to SurveyMonkey. If you want to create a short survey (10 questions) and send it to small groups (100 and under) of users, you can use SurveyMonkey for free. Additional options, such as unlimited questions and users or data export, cost extra. SurveyMonkey will even help you with audience targeting if you have the need and the budget.

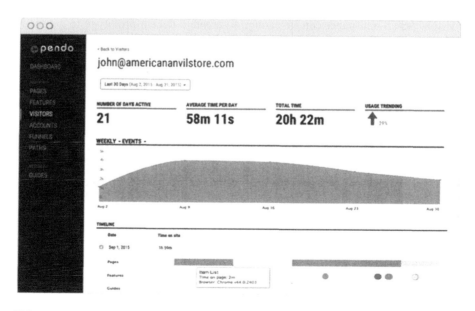

FIGURE 15.1   Pendo user analytics platform. (Reprinted with permission from Pendo.)

SurveyMonkey is a reliable, flexible solution. You can modify built-in question types, such as multiple choice and rating scales. Should you not have the time or expertise to create a custom survey, you also can make use of hundreds of free survey questionnaires already created and vetted for validity. There's a reason why SurveyMonkey is in such wide use—it works.

15.1.3 OTHER USER RESEARCH TOOLS

Simple Card Sort: Although there is, at minimum, a $49.99 price for 30 days of use, SimpleCardSort enables you to quickly create a card-sorting activity and invite participants to complete it. Participants can do this activity anywhere, and once they're done, you get the results, viewable in different formats, individually and collectively.

Optimal Workshop: More robust than SimpleCardSort is Optimal Workshop. It offers card sorting as well as tree testing (which is often called reverse card sorting), first click analysis, and online usability testing and analysis. Prices range from $82 to $166 a month although you can purchase one survey for $149. If you want to see and understand your users from multiple perspectives and have money to support this, Optimal Workshop offers considerable support, both in testing as well as subsequent data analysis of test results.

15.2 PRODUCT PROTOTYPING TOOLS

15.2.1 AXURE

We have used Axure extensively because of its easy learning curve, support community, and capability to support the fast development of applications, especially those deployed for mobile devices. Brian, in fact, teaches regularly with it in his undergraduate web and interaction design classes because students can see results almost immediately. After a short, online series of tutorial videos, students are dragging and dropping objects, including buttons and images, and giving them interactivity. They then can publish their work and view it in the environment in which it will be used, such as a smartphone (Figure 15.2).

The standard version of the app costs $289 per license, but students and teachers can get 60% off that price. Axure allows for fast prototyping, but it also can be used to create a finished, higher-fidelity product, and that flexibility along with ease of use make it a tool you should consider for your UCD work.

15.2.2 PENCIL PROJECT

Free to use, Pencil is a great mock-up tool for low-fidelity prototyping. It works on any platform (i.e., Mac, Windows) and comes with an expansive library of widgets, shapes, and other features that allow you to work quickly to develop a prototype, complete with interactivity, for users to test (Figure 15.3).

FIGURE 15.2 Axure prototyping software. (Reprinted with permission from Axure.)

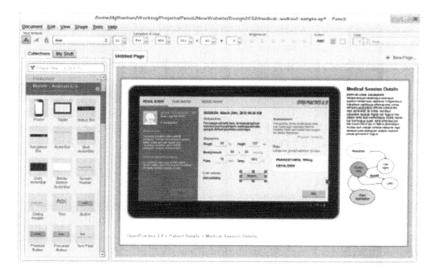

FIGURE 15.3 Pencil mockup tool. (Reprinted with permission from Pencil Project.)

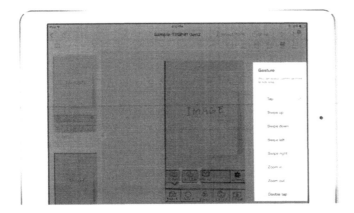

FIGURE 15.4 POP application. (Reprinted with permission from POP.)

15.2.3 POP

We first came across prototyping on paper (POP) when one of Brian's graduate students needed a way to have users test her paper prototype but couldn't be on location where the student resided. The POP app allows you to very quickly put your paper prototype into a digital environment, giving it interactive features, including gestures, such as swiping, touching, or pinching, without having to render your paper prototype into a more refined fidelity (Figure 15.4). The basic version of POP is free and available for iOS, Android, and Windows mobile platforms.

15.2.4 OTHER PROTOTYPING TOOLS

Invision: Let's say you already have a more refined design and want to prototype it. Invision, which is free, allows you to create interactivity for your design, turning it into a prototype that can be viewed and used on desktops, laptops, and mobile devices. It also has, in its enterprise version, a project management tool.

Balsamiq: Another mock-up tool for low- and medium-fidelity sketches and wireframes, Balsamiq, under $100 for a single user license, is a more established application with a comprehensive set of editing features. We have used Balsamiq and like its capability to be used relatively quickly and easily to create prototypes.

Justinmind: Many professional developers use Justinmind. It has garnered multiple awards, in fact, because of its quality support for the development of high-fidelity prototypes. You can even create apps for Google Glass with Justinmind. At $19 a month per user, it's affordable, and that low price pays for many things, including visual design, mobile gestures, image hotspots, web and mobile simulation along with user testing and even team management through version control and publication management to include feedback from users. If you are going to build an app that is core to your business and want every capability and feature, especially if you have a team of developers, Justinmind is definitely a prototyping tool to consider.

15.3   USER TESTING TOOLS

15.3.1   TECHSMITH MORAE

You don't need to have usability testing and analysis software in every instance if you're developing a product. But if you are in a situation in which you need to collect data, analyze it quantitatively and qualitatively, and then report on what you've learned in clear, all compelling ways to convince key stakeholders of your decisions, TechSmith's Morae app is likely a good choice for you to consider.

It's expensive (around $1500 for education and $2000 for noneducation licensing), but you can get a free 15-day trial. You buy a great deal with that, including the ability to run usability tests that record user video/audio, many user interactions with a product (such as keystrokes, mouse clicks, usage time), and also markings, all of which can be scored, from observers during testing or after the fact. Additionally, Morae comes with built-in data crunching capability, including determining standard deviations of user responses. Morae allows you to report data through standard or custom graphs and tables that can then be exported directly into a PowerPoint presentation along with video/audio clips. You can see all of this in action in Chapter 9. Most of the report slide examples were generated, in part, using Morae.

Our lab has used Morae quite extensively over the last few years because of the heavy reporting requirements of many of our client contracts. However, it is limited. You can't use it on Mac, and to test mobile devices, you need another web camera to record the screen of the device because Morae won't run in that environment. Remote testing of users can also be a challenge. UserTesting is a better alternative application if you have to go that route.

15.3.2   SILVERBACK

The latest version of Silverback is under $100, and older versions, supporting older systems, are now free shareware. Silverback is "guerilla" usability testing. It has enough features to allow you to create and record users performing tasks. Like Morae, it also records user screen interactions as well as their video/audio. Unlike Morae, Silverback does work on Mac.

15.3.3   USERTESTING

One of the drawbacks to user testing is that often your population is pretty convenient. You recruit people who live or work around you and who are easiest to get into your location to try out your product. But just because they are convenient to recruit doesn't mean they are representative of all of your users. In fact, the more convenient the sample, the greater danger there is that you aren't getting feedback from the right users.

One of the benefits of UserTesting is that the service will recruit your users, even create a test plan to try out your product, and will then record users' video/audio, written comments, and any results of task performance when using your product, returning all of this data very quickly. Plus, because of its reach, UserTesting can track down for you users who match your profile anywhere in the world.

Now this isn't cheap. You might get a few free trials, but, over time, you should expect to pay in the range of $50 per user result. If you're doing this all on your own or have a very limited budget, that's too much to spend. You can do other things on your own to carry out at least qualitative testing and analysis. But if you have money, and if this product is a big deal for your client or organization, $50 per user or $1500 for a lot of users, well, you get what you pay for. If you are working in a Windows environment, if you conduct a lot of focus groups, if you know you're going to do a lot more testing of other products, and if you have the skill to do testing and analysis, consider Morae. If you want to test in a mobile setting and want someone else to handle all the testing and reporting for you, UserTesting may be the best route to take.

This list isn't exhaustive, and the absence of a tool here doesn't imply that tool is lacking. We either didn't have the space to include it, or we lacked a familiarity with how it works to comment intelligently on how it could be useful for you. By all means, experiment to find which tools work best for you, whether they are listed here or not. The right tool(s) for you accommodate your budget, your skill level, and your product scope (time, materials, and environmental demands).

Index

Page numbers followed by f and t indicate figures and tables, respectively.